HARCOURT Science

Harcourt School Publishers

Orlando • Boston • Dallas • Chicago • San Diego

www.harcourtschool.com

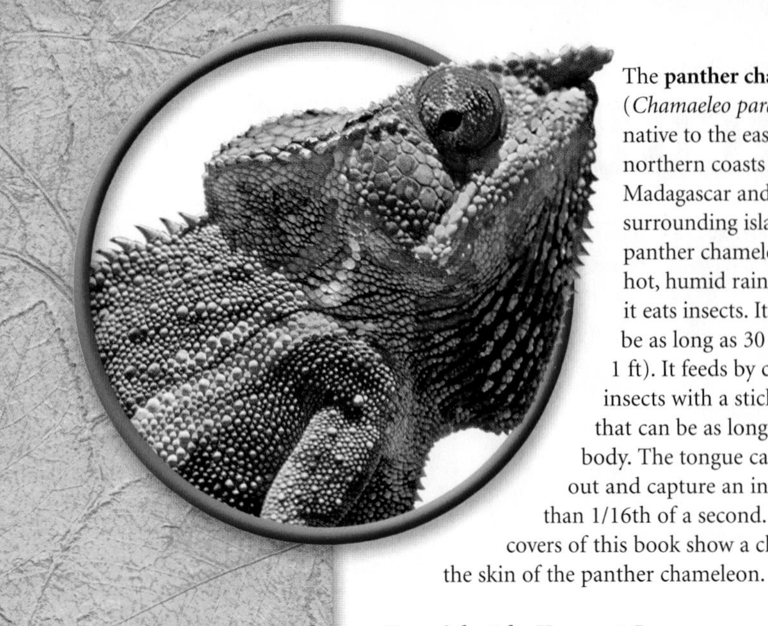

The **panther chameleon** (*Chamaeleo pardalis*) is native to the eastern and northern coasts of Madagascar and some surrounding islands. The panther chameleon lives in hot, humid rain forests and it eats insects. It can grow to be as long as 30 cm (about 1 ft). It feeds by capturing insects with a sticky tongue that can be as long as its entire body. The tongue can extend out and capture an insect in less than 1/16th of a second. The inside covers of this book show a closeup of the skin of the panther chameleon.

2005 Edition

Printed in the United States of America

ISBN 0-15-325389-4 UNIT A
ISBN 0-15-325390-8 UNIT B
ISBN 0-15-325391-6 UNIT C
ISBN 0-15-325392-4 UNIT D
ISBN 0-15-325393-2 UNIT E
ISBN 0-15-325394-0 UNIT F

8 9 10 11 12 13 032 10 09 08 07

Authors

Marjorie Slavick Frank
Former Adjunct Faculty Member
Hunter, Brooklyn, and Manhattan Colleges
New York, New York

Robert M. Jones
Professor of Education
University of Houston–Clear Lake
Houston, Texas

Gerald H. Krockover
Professor of Earth and Atmospheric Science Education
School Mathematics and Science Center
Purdue University
West Lafayette, Indiana

Mozell P. Lang
Science Education Consultant
Michigan Department of Education
Lansing, Michigan

Joyce C. McLeod
Visiting Professor
Rollins College
Winter Park, Florida

Carol J. Valenta
Vice President—Education, Exhibits, and Programs
St. Louis Science Center
St. Louis, Missouri

Barry A. Van Deman
Program Director, Informal Science Education
Arlington, Virginia

Senior Editorial Advisor

Napoleon Adebola Bryant, Jr.
Professor Emeritus of Education
Xavier University
Cincinnati, Ohio

Program Advisors

Michael J. Bell
Assistant Professor of Early Childhood Education
School of Education
West Chester University
West Chester, Pennsylvania

George W. Bright
Professor of Mathematics Education
The University of North Carolina at Greensboro
Greensboro, North Carolina

Pansy Cowder
Science Specialist
Tampa, Florida

Robert H. Fronk
Head, Science/Mathematics Education Department
Florida Institute of Technology
Melbourne, Florida

Bernard A. Harris, Jr.
Physician and Former Astronaut
(STS 55–Space Shuttle *Columbia*, STS 63–Space Shuttle *Discovery*)
President, The Harris Foundation
Houston, Texas

Lois Harrison-Jones
Education and Management Consultant
Dallas, Texas

Kenneth R. Mechling
Professor of Biology and Science Education
Clarion University of Pennsylvania
Clarion, Pennsylvania

Nancy Roser
Professor of Language and Literacy Studies
University of Texas, Austin
Austin, Texas

Program Advisor and Activities Writer

Barbara ten Brink
Science Director
Round Rock Independent School District
Round Rock, Texas

Reviewers and Contributors

Rose G. Baublitz
Director, Curriculum and Staff Development
Granville Exempted Village School District
Granville, Ohio

Sharon W. Bowers
Teacher
Kemps Landing Magnet School
Virginia Beach, Virginia

Joe E. Hart
Science Lead Teacher
Clayton County Public Schools
Morrow, Georgia

Kathryn Henry
Teacher Trainer
Public School CS 200
New York, New York

Denise Hunt
Teacher
Cumming Elementary School
Cumming, Georgia

Marie A. McDermott
Teacher
Kingsbury School
Waterbury, Connecticut

Clyde Partner
Science/Health Curriculum Coordinator
Evanston S.D. #65
Evanston, Illinois

Michael F. Ryan
Educational Technology Specialist
Lake County Schools
Tavares, Florida

Arnold E. Serotsky
Teacher/Science Department Coordinator
Greece Athena Middle School
Rochester, New York

Lynda Wood
Science Coordinator
Southfield Education Center
Southfield, Michigan

UNIT A LIFE SCIENCE

Living Systems

UNIT B LIFE SCIENCE

Systems and Interactions in Nature

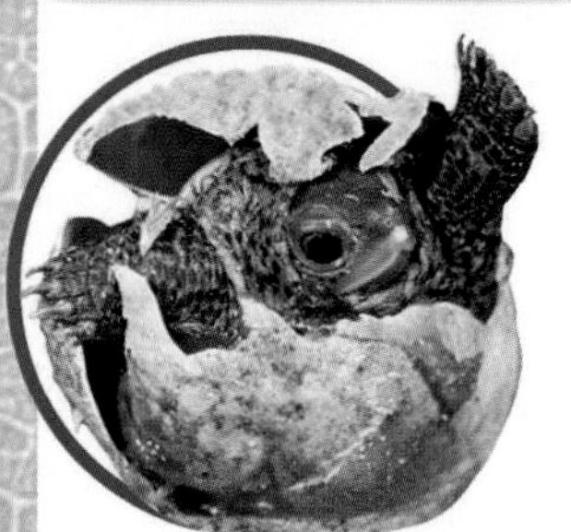

UNIT C

EARTH SCIENCE

Processes That Change the Earth

UNIT D

EARTH SCIENCE
The Solar System and Beyond

UNIT E

PHYSICAL SCIENCE

Building Blocks of Matter

UNIT F

PHYSICAL SCIENCE

Energy and Motion

Planning an Investigation

How do scientists answer a question or solve a problem they have identified? They use organized ways called **scientific methods** to plan and conduct a study. They use science process skills to help them gather, organize, analyze, and present their information.

Justin is using this scientific method for experimenting to find an answer to his question. You can use these steps, too.

STEP 1 Observe and ask questions.

- Use your senses to make observations.
- Record **one** question that you would like to answer.
- Write down what you already know about the topic of your question.
- Decide what other information you need.
- Do research to find more information about your topic.

What design of paper airplane will fly the greatest distance? I need to find out more about airplane wings.

STEP 2 Form a hypothesis.

- Write a possible answer to your question. A possible answer to a question that can be tested is a **hypothesis**.
- Write your hypothesis in a complete sentence.

STEP 3 Plan an experiment.

- Decide how to conduct a fair test of your hypothesis by controlling variables. **Variables** are factors that can affect the outcome of the investigation.
- Write down the steps you will follow to do your test.
- List the equipment you will need.
- Decide how you will gather and record your data.

I'll launch each airplane three times. Each airplane will be launched from the same spot, and I'll use the same amount of force each time.
C
B
A

STEP 4 Conduct the experiment.

- Follow the steps you wrote.
- Observe and measure carefully.
- Record everything that happens.
- Organize your data so you can study it carefully.

I'll record each distance. Then I'll find the average distance each airplane traveled.

STEP 5 Draw conclusions and communicate results.

- Analyze the data you gathered.
- Make charts, tables, or graphs to show your data.
- Write a conclusion. Describe the evidence you used to determine whether your test supported your hypothesis.
- Decide whether your hypothesis was correct.

My hypothesis was correct. The airplane with the narrow wings flew farthest.

INVESTIGATE FURTHER

What if your hypothesis was correct . . .

You may want to pose another question about your topic that you can test.

What if your hypothesis was incorrect . . .

You may want to form another hypothesis and do a test on a different variable.

I'll test this new hypothesis: The airplane with the narrow wings will also fly for the longest time.

Do you think Justin's new hypothesis will be correct? Plan and conduct a test to find out!

Using Science Process Skills

When scientists try to find an answer to a question or do an experiment, they use thinking tools called **process skills**. You use many of the process skills whenever you speak, listen, read, write, or think. Think about how these students use process skills to help them answer questions, do experiments, and investigate the world around them.

What Greg plans to investigate

Greg is finding leaves in the park. He wants to make collections of leaves that are alike in some way. He looks for leaves of different sizes and shapes.

Process Skills

Observe—use the senses to learn about objects and events

Compare—identify characteristics about things or events to find out how they are alike and different

Measure—compare an attribute of an object, such as its mass, length, or volume, to a standard unit, such as a gram, centimeter, or liter

Classify—group or organize objects or events in categories based on specific characteristics

How Greg uses process skills

He **observes** the leaves and **compares** their sizes, shapes, and colors. He **measures** each leaf with a ruler. Then he **classifies** the leaves, first into groups based on their sizes and then into groups based on their shapes.

What Pilar plans to investigate

It's been raining for part of the week. Pilar wants to know if it will rain during the coming weekend.

How Pilar uses process skills

She **gathers and records data** to make a prediction about the weather. She observes the weather each day of the week and records it. On a chart, she **displays data** she has gathered. On Friday, she **predicts**, based on her observations, that it will rain during the weekend.

Process Skills

Gather, Record, Display, or Interpret Data

- gather data by making observations and use them to make inferences or predictions
- record data by writing down observations
- display data by making tables, charts, or graphs

Predict—form an idea of an expected outcome, based on observations or experience

Process Skills

Hypothesize—make a statement about an expected outcome, based on observation, knowledge, and experience

Plan and Conduct a Simple Investigation—identify and perform the steps necessary to find the answer to a question, using appropriate tools and recording and analyzing the data collected

Infer—use logical reasoning to explain events and draw conclusions based on observations

What Tran plans to investigate

Tran is interested in knowing how the size of a magnet is related to its strength.

How Tran uses process skills

He **hypothesizes** that larger magnets are stronger than smaller magnets. He **plans and conducts a simple investigation** to see if his hypothesis is correct. He gathers magnets of different sizes and objects of different weights that the magnets will attract. Tran tests each item with each magnet and records his findings. His hypothesis seems to be correct until he tests the last object, a toy truck. When the large bar magnet cannot pick up the truck, but the smaller horseshoe magnet can, he **infers** that the largest magnets are not always the strongest.

What Emily plans to investigate

Emily sees an ad about food wrap. The people in the ad claim that Tight-Right food wrap seals containers better than other food wraps. Emily plans a simple experiment to find out if this claim is true.

How Emily uses process skills

She **identifies and controls variables** in the experiment by choosing three bowls that are exactly the same. She labels the bowls A, B, and C, places them on a tray, and adds exactly 350 mL of water to each bowl. She cuts a 25-cm-long piece of Tight-Right food wrap and covers bowl A. She cuts 25-cm-long pieces of two other brands of food wrap and covers bowls B and C. She seals the food wrap on all three bowls as tightly as she can. Emily **experiments** with the seals by shaking the tray on which the bowls are placed. Water sloshes up the sides of the bowls and leaks out onto the tray from bowls B and C, but not from bowl A. From her observations she infers that the claim for Tight-Right food wrap is true.

Process Skills

Identify and Control Variables—identify and control factors that affect the outcome of an experiment

Experiment—design ways to collect data to test hypotheses under controlled conditions

Reading to Learn

Scientists use reading, writing, and numbers in their work. They **read** to find out everything they can about a topic they are investigating. So it is important that scientists know the meanings of science vocabulary and that they understand what they read. Use the following strategies to help you become a good science reader!

Before Reading

- Read the Find Out statements to help you know what to look for as you read.
- Think: I need to find out how living things get the energy they need.

- Look at the Vocabulary terms.
- Be sure you can pronounce each term.
- Look up each term in the Glossary.
- Say the definition to yourself. Use the term in a sentence to show its meaning.

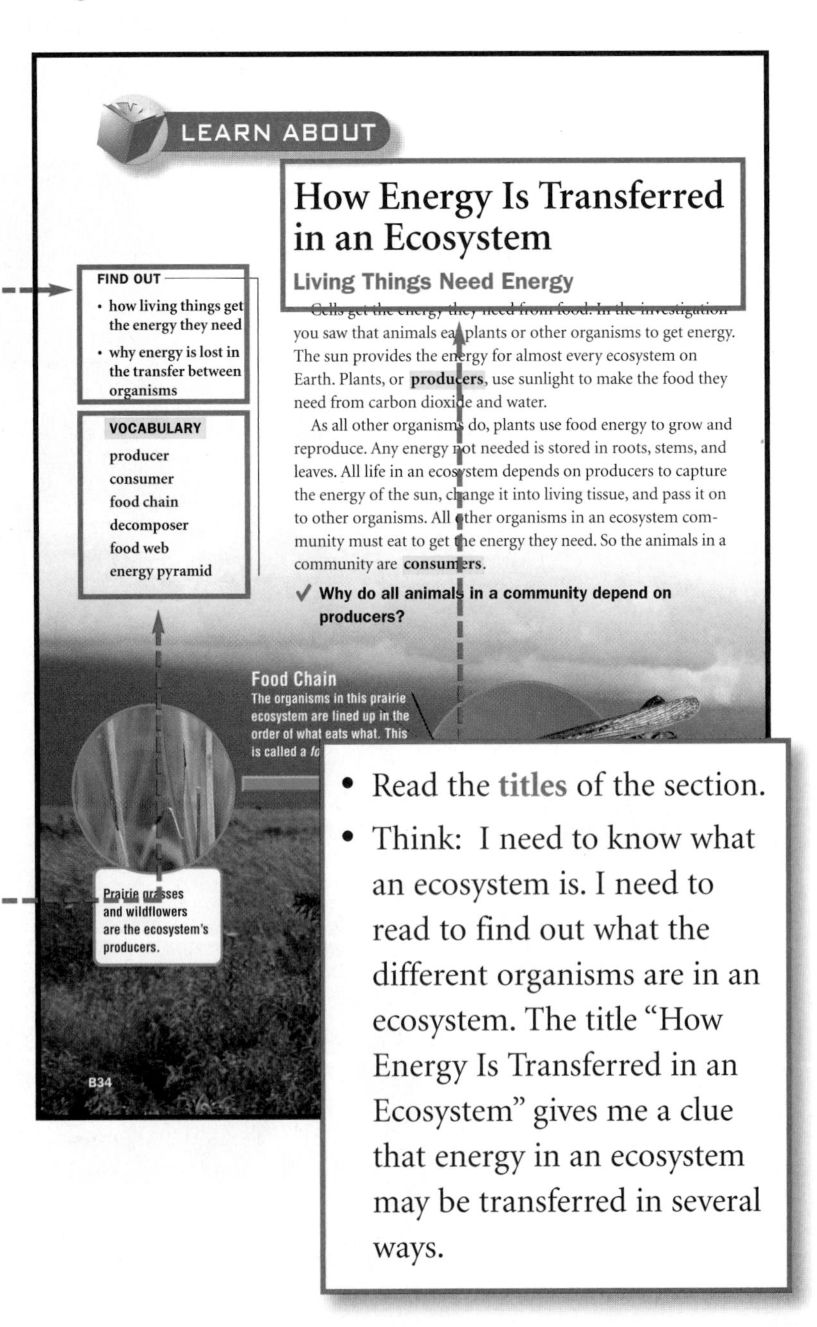

LEARN ABOUT
How Energy Is Transferred in an Ecosystem
Living Things Need Energy
FIND OUT
• how living things get the energy they need
• why energy is lost in the transfer between organisms
VOCABULARY
producer
consumer
food chain
decomposer
food web
energy pyramid
you saw that animals eat plants or other organisms to get energy. The sun provides the energy for almost every ecosystem on Earth. Plants, or producers, use sunlight to make the food they need from carbon dioxide and water.
As all other organisms do, plants use food energy to grow and reproduce. Any energy not needed is stored in roots, stems, and leaves. All life in an ecosystem depends on producers to capture the energy of the sun, change it into living tissue, and pass it on to other organisms. All other organisms in an ecosystem community must eat to get the energy they need. So the animals in a community are consumers.
Why do all animals in a community depend on producers?
Food Chain
The organisms in this prairie ecosystem are lined up in the order of what eats what. This
Prairie grasses and wildflowers are the ecosystem's producers.
B34

- Read the titles of the section.
- Think: I need to know what an ecosystem is. I need to read to find out what the different organisms are in an ecosystem. The title "How Energy Is Transferred in an Ecosystem" gives me a clue that energy in an ecosystem may be transferred in several ways.

During Reading

Find the **main idea** in the first paragraph.

- In food chains, there are more producers than consumers.

Find the **details** in the next paragraph that support the main idea.

- Only 10 percent of the energy at any level is passed on to the next level.
- High-level consumers, such as wolves, have relatively small populations. There isn't enough energy for large populations of wolves.
- All other organisms in an ecosystem must eat to get the energy they need.

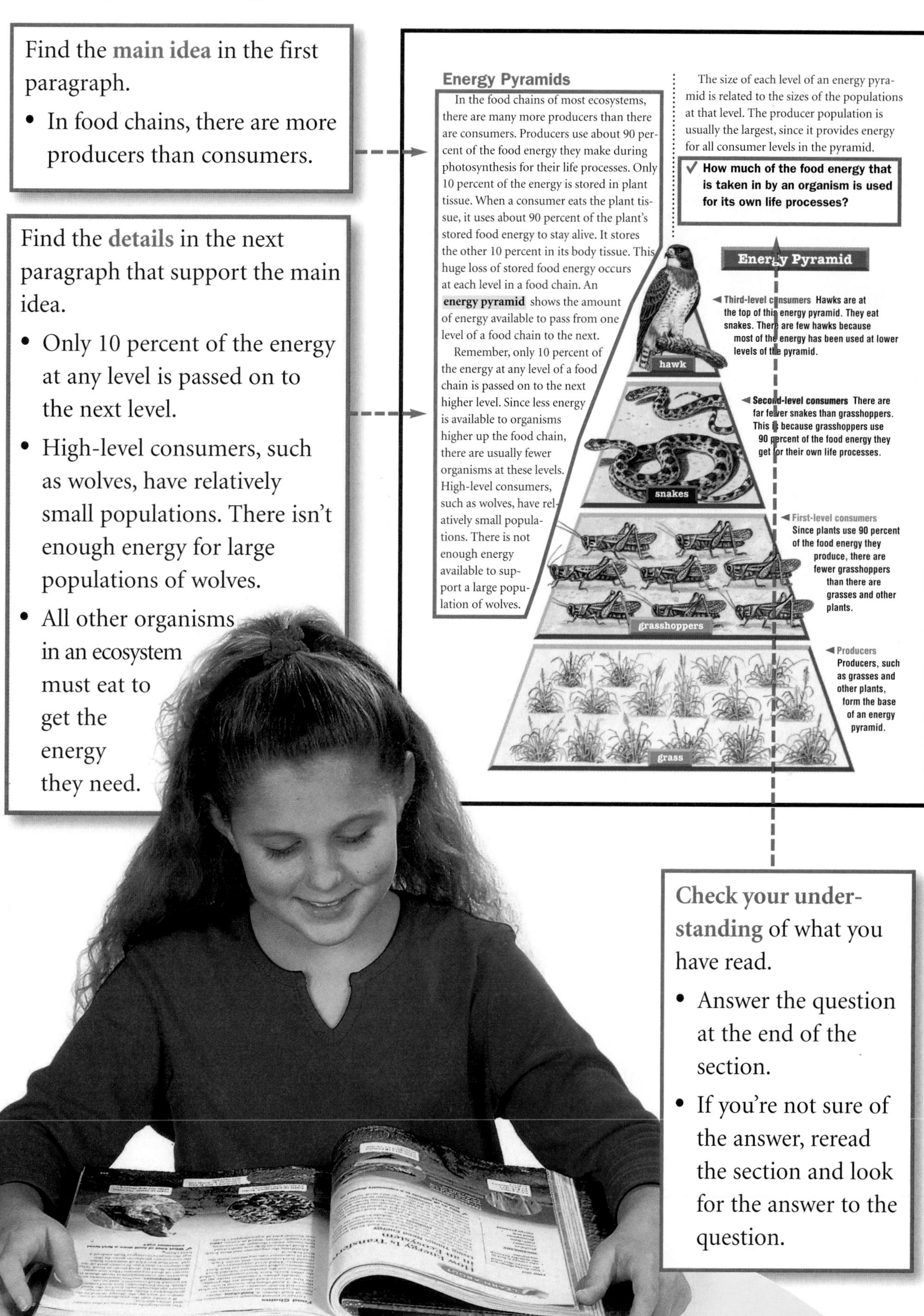

Energy Pyramids

In the food chains of most ecosystems, there are many more producers than there are consumers. Producers use about 90 percent of the food energy they make during photosynthesis for their life processes. Only 10 percent of the energy is stored in plant tissue. When a consumer eats the plant tissue, it uses about 90 percent of the plant's stored food energy to stay alive. It stores the other 10 percent in its body tissue. This huge loss of stored food energy occurs at each level in a food chain. An **energy pyramid** shows the amount of energy available to pass from one level of a food chain to the next.

Remember, only 10 percent of the energy at any level of a food chain is passed on to the next higher level. Since less energy is available to organisms higher up the food chain, there are usually fewer organisms at these levels. High-level consumers, such as wolves, have relatively small populations. There is not enough energy available to support a large population of wolves.

The size of each level of an energy pyramid is related to the sizes of the populations at that level. The producer population is usually the largest, since it provides energy for all consumer levels in the pyramid.

✓ **How much of the food energy that is taken in by an organism is used for its own life processes?**

Energy Pyramid
Third-level consumers Hawks are at the top of this energy pyramid. They eat snakes. There are few hawks because most of the energy has been used at lower levels of the pyramid.
hawk
Second-level consumers There are far fewer snakes than grasshoppers. This is because grasshoppers use 90 percent of the food energy they get for their own life processes.
snakes
First-level consumers Since plants use 90 percent of the food energy they produce, there are fewer grasshoppers than there are grasses and other plants.
grasshoppers
Producers Producers, such as grasses and other plants, form the base of an energy pyramid.
grass

Check your **understanding** of what you have read.

- Answer the question at the end of the section.
- If you're not sure of the answer, reread the section and look for the answer to the question.

After Reading

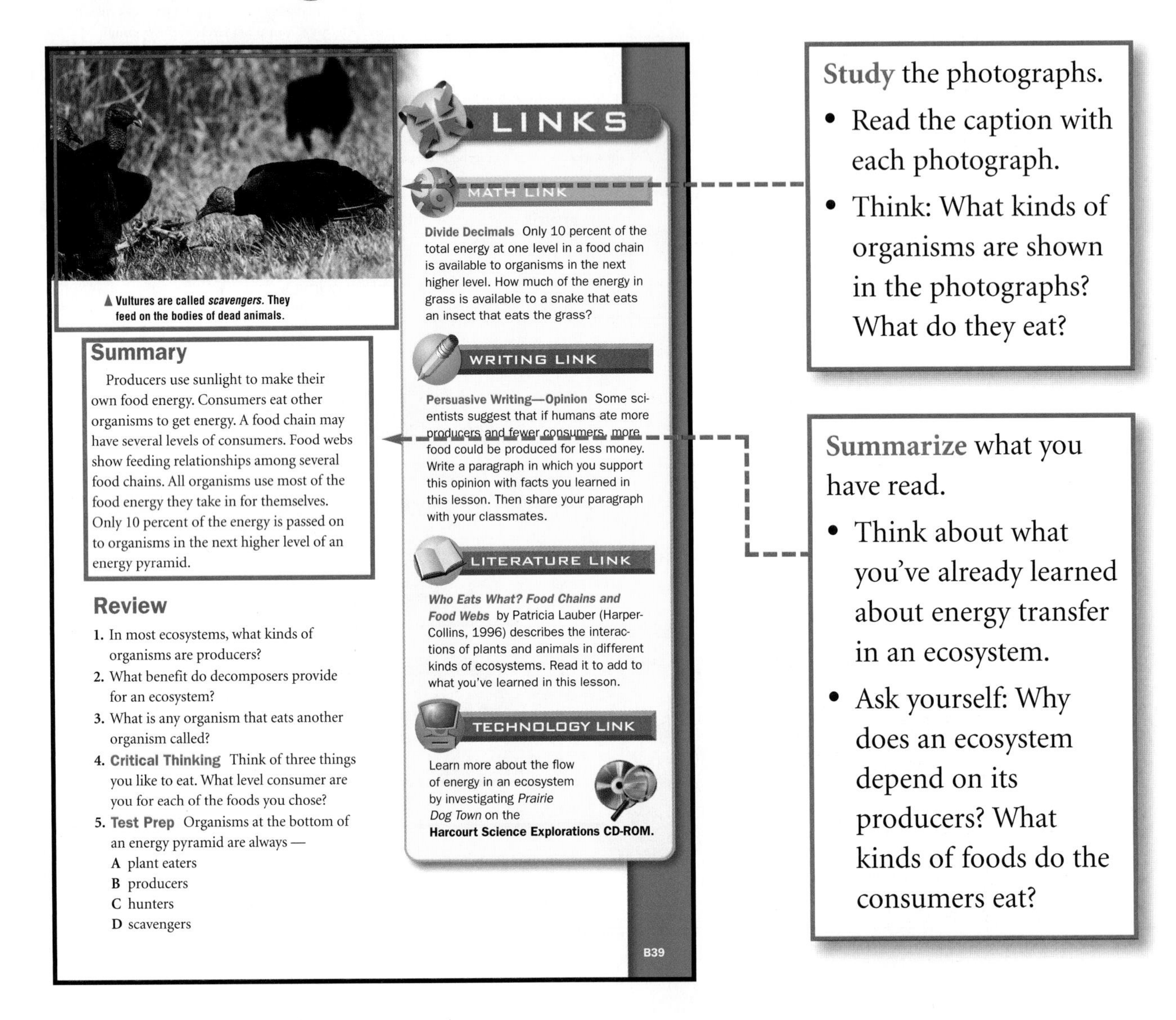

▲ Vultures are called *scavengers*. They feed on the bodies of dead animals.
LINKS
MATH LINK
Divide Decimals Only 10 percent of the total energy at one level in a food chain is available to organisms in the next higher level. How much of the energy in grass is available to a snake that eats an insect that eats the grass?
Summary
Producers use sunlight to make their own food energy. Consumers eat other organisms to get energy. A food chain may have several levels of consumers. Food webs show feeding relationships among several food chains. All organisms use most of the food energy they take in for themselves. Only 10 percent of the energy is passed on to organisms in the next higher level of an energy pyramid.
WRITING LINK
Persuasive Writing—Opinion Some scientists suggest that if humans ate more producers and fewer consumers, more food could be produced for less money. Write a paragraph in which you support this opinion with facts you learned in this lesson. Then share your paragraph with your classmates.
Review
1. In most ecosystems, what kinds of organisms are producers?
2. What benefit do decomposers provide for an ecosystem?
3. What is any organism that eats another organism called?
4. **Critical Thinking** Think of three things you like to eat. What level consumer are you for each of the foods you chose?
5. **Test Prep** Organisms at the bottom of an energy pyramid are always —
A plant eaters
B producers
C hunters
D scavengers
LITERATURE LINK
Who Eats What? Food Chains and Food Webs by Patricia Lauber (HarperCollins, 1996) describes the interactions of plants and animals in different kinds of ecosystems. Read it to add to what you've learned in this lesson.
TECHNOLOGY LINK
Learn more about the flow of energy in an ecosystem by investigating *Prairie Dog Town* on the **Harcourt Science Explorations CD-ROM.**
B39

Study the photographs.

- Read the caption with each photograph.
- Think: What kinds of organisms are shown in the photographs? What do they eat?

Summarize what you have read.

- Think about what you've already learned about energy transfer in an ecosystem.
- Ask yourself: Why does an ecosystem depend on its producers? What kinds of foods do the consumers eat?

For more reading strategies and tips, see pages R38–R49.

Reading about science helps you understand your conclusions from your investigations.

Writing to Communicate

Writing about what you are learning helps you connect the new ideas to what you already know. Scientists **write** about what they learn in their research and investigations to help others understand the work they have done. As you work like a scientist, you will use the following kinds of writing to describe what you are doing and learning.

In **informative writing**, you may

- describe your observations, inferences, and conclusions.
- tell how to do an experiment.

In **narrative writing**, you may

- describe something, give examples, or tell a story.

In **expressive writing**, you may

- write letters, poems, or songs.

In **persuasive writing**, you may

- write letters about important issues in science.
- write essays expressing your opinions about science issues.

Writing about what you have learned about science helps others understand your thinking.

Using Numbers

Scientists **use numbers** when they collect, display, and interpret their data. Understanding numbers and using them correctly to show the results of investigations are important skills that a scientist must have. As you work like a scientist, you will use numbers in the following ways:

Measuring

Scientists make accurate measurements as they gather data. They use measuring instruments such as thermometers, clocks and timers, rulers, a spring scale, and a balance, and they use beakers and other containers to measure liquids.

For more information about using measuring tools, see pages R2–R6.

Interpreting Data

Scientists collect, organize, display, and interpret data as they do investigations. Scientists choose a way to display data that helps others understand what they have learned. Tables, charts, and graphs are good ways to display data so that it can be interpreted by others.

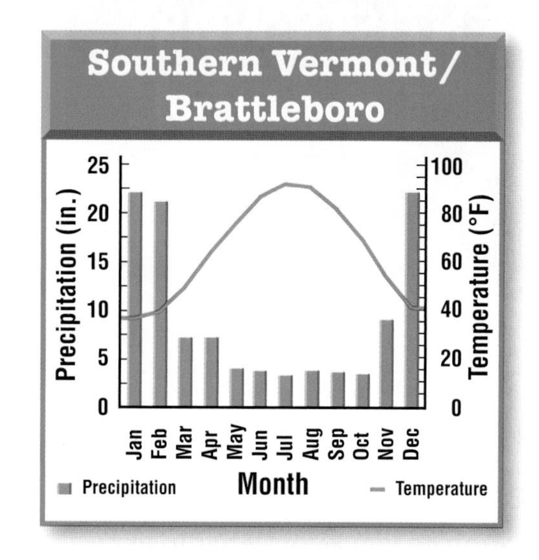

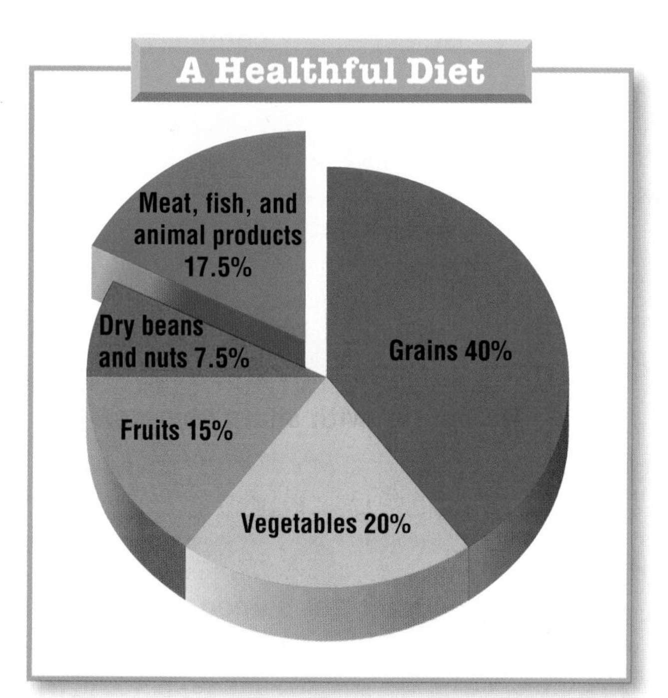

Using Number Sense

Scientists must understand what the numbers they use represent. They compare and order numbers, compute with numbers, read and understand the numbers shown on graphs, and read the scales on thermometers, measuring cups, beakers, and other tools.

Good scientists apply their math skills to help them display and interpret the data they collect.

In *Harcourt Science* you will have many opportunities to work like a scientist. An exciting year of discovery lies ahead!

Safety in Science

Doing investigations in science can be fun, but you need to be sure you do them safely. Here are some rules to follow.

1. **Think ahead.** Study the steps of the investigation so you know what to expect. If you have any questions, ask your teacher. Be sure you understand any safety symbols that are shown.
2. **Be neat.** Keep your work area clean. If you have long hair, pull it back so it doesn't get in the way. Roll or push up long sleeves to keep them away from your experiment.
3. **Oops!** If you should spill or break something, or get cut, tell your teacher right away.
4. **Watch your eyes.** Wear safety goggles anytime you are directed to do so. If you get anything in your eyes, tell your teacher right away.
5. **Yuck!** Never eat or drink anything during a science activity.
6. **Don't get shocked.** Be especially careful if an electric appliance is used. Be sure that electric cords are in a safe place where you can't trip over them. Don't ever pull a plug out of an outlet by pulling on the cord.
7. **Keep it clean.** Always clean up when you have finished. Put everything away and wipe your work area. Wash your hands.

In some activities you will see these symbols. They are signs for what you need to be safe.

Be especially careful.

Wear safety goggles.

Be careful with sharp objects.

Don't get burned.

Protect your clothes.

Protect your hands with mitts.

Be careful with electricity.

UNIT D EARTH SCIENCE

The Solar System and Beyond

UNIT EXPERIMENT

Designing Rockets

Scientists study the universe from Earth and from space. Studying the universe from space requires rockets that can send heavy payloads deep into space. While you study this unit, you can conduct a long-term experiment on rocket design. Here are some questions to think about. What rocket design will carry heavy payloads the farthest? For example, is one large rocket more powerful than several smaller rockets? Plan and conduct an experiment to find answers to these or other questions you have about designing rockets. See pages x–xvii for help in designing your experiment.

CHAPTER

1

Earth, Moon, and Beyond

Vocabulary Preview

revolve
orbit
rotate
axis
eclipse
solstice
equinox
planets
asteroids
comets
telescope
satellite
space probe

Although Earth moves through space very rapidly, it doesn't *seem* to move at all. That's why the sun and the moon both appear to move around Earth. It took astronomers a long time to realize that Earth actually moves around the sun.

Fast Fact

Observatories located near growing cities are no longer very useful. The city lights and the pollution in the air make it impossible to see the stars clearly. But on the summit of Mauna Kea in Hawai'i, many miles from city lights and pollution, scientists can see the stars quite clearly.

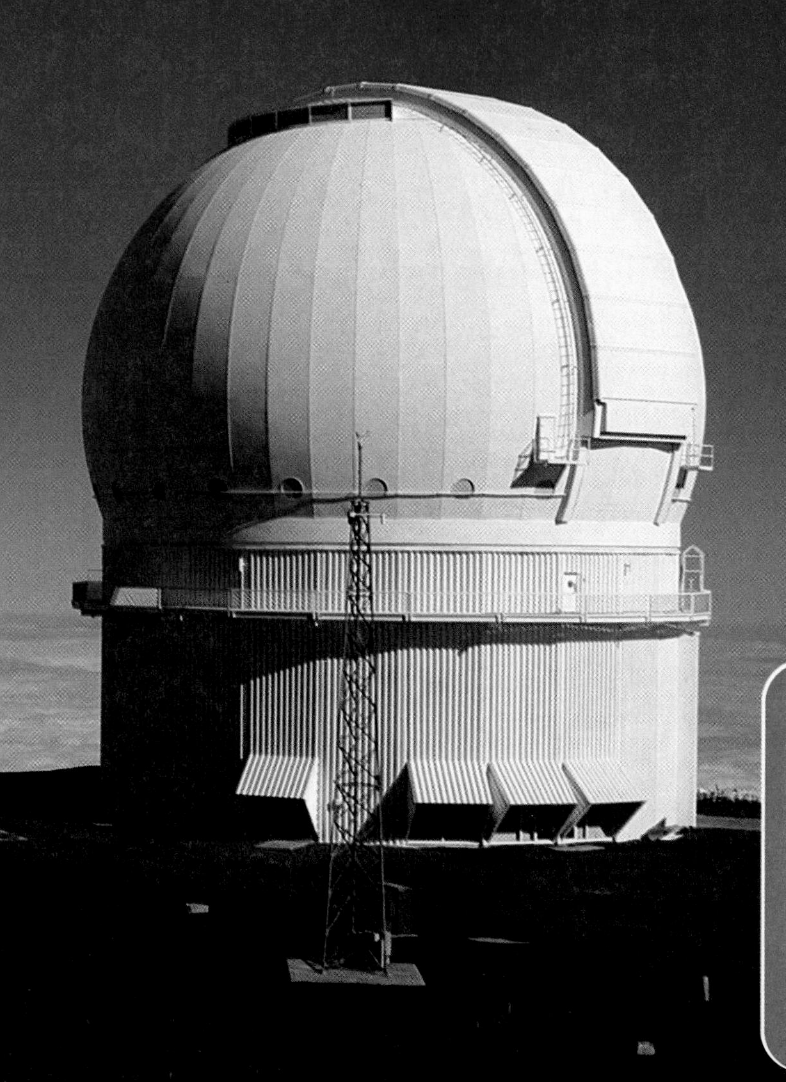

Fast Fact

It takes Earth 365.26 days to travel once around the sun. An adjustment to the calendar is made for the 0.26 day by adding a February 29 every four years. These longer years are called leap years.

Once Around the Sun

Planet	Earth Days
Mercury	88
Venus	224.7
Earth	365.26
Mars	687
Jupiter	4332.6
Saturn	10,759.2
Uranus	30,685.4
Neptune	60,189
Pluto	90,777.6

Fast Fact

The next total solar eclipse in the United States will take place on August 21, 2017. It will move across the country from Oregon to South Carolina. You can safely view a solar eclipse the way the photograph shows.

LESSON 1

How Do the Earth and the Moon Compare?

In this lesson, you can . . .

INVESTIGATE how the moon craters were formed.

LEARN ABOUT Earth and its moon.

LINK to math, writing, music, and technology.

Earth, from the surface of the moon ▼

The Moon's Craters

Activity Purpose As Earth's nearest neighbor in space, the moon was the first object in the solar system that people studied. People **observed** that the moon's surface was very different from the Earth's surface. One difference was the large number of craters on the moon. In the investigation you will **make a model** of the moon's surface to **infer** how the craters formed.

Materials

- newspaper
- aluminum pan
- large spoon
- $\frac{1}{2}$cup water
- 1 cup flour
- safety goggles
- apron
- marble
- meterstick

Activity Procedure

1. Copy the table below.

Trial	Height	Width of Craters
1	20 cm	
2	40 cm	
3	80 cm	
4	100 cm	

2. Put the newspaper on the floor. Place the pan in the center of the newspaper.
3. Use a large spoon to mix the water and flour in the aluminum pan. The look and feel of the mixture should be like thick cake batter. Now lightly cover the surface of the mixture with dry flour. (Picture A)

4 CAUTION **Put on the safety goggles and apron** to protect your eyes and clothes from flour dust. Drop the marble into the pan from a height of 20 cm. (Picture B)

5 Carefully remove the marble and **measure** the width of the crater. **Record** the measurement in the table. Repeat Steps 4 and 5 two more times.

6 Now drop the marble three times each from heights of 40 cm, 80 cm, and 100 cm. **Measure** the craters and **record** the measurements after each drop.

Picture A

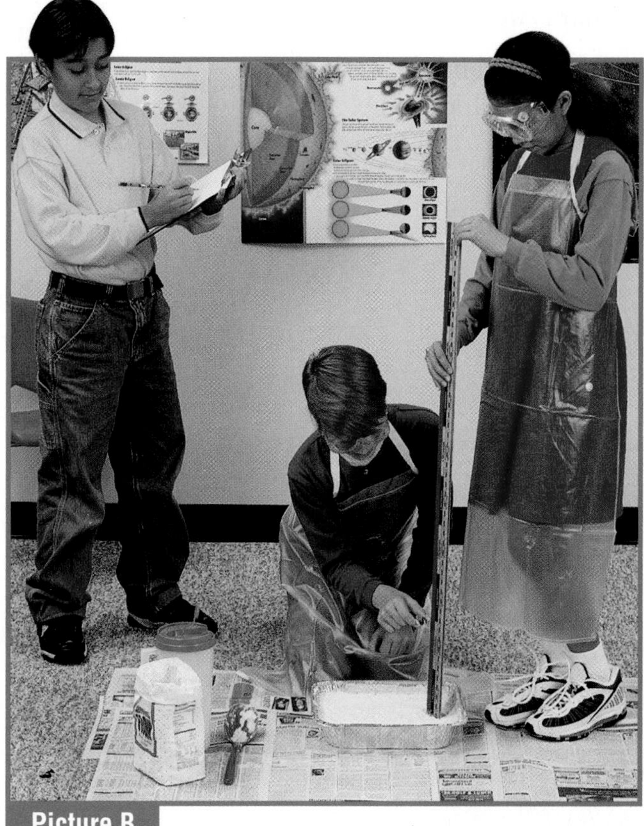
Picture B

Draw Conclusions

1. **Compare** the height from which each marble was dropped to the size of the crater it made. How does height affect crater size?
2. The Copernicus (koh•PER•nih•kuhs) crater on the moon is 91 km across. Based on your model, what can you **infer** about the object that formed this crater?
3. **Scientists at Work** Most of the moon's craters were formed millions of years ago. Scientists **use models** to **infer** events that occurred too long ago to **observe** directly. What did you infer from the model about how the moon's craters formed?

Investigate Further **Hypothesize** how using larger or smaller marbles would affect the size and shape of the craters. **Plan and conduct a simple experiment** to test your hypothesis.

Process Skill Tip

You can **use a model** to **infer** how something happened a long time ago, such as how the moon's craters formed.

How Earth and the Moon Compare

FIND OUT

- about the Earth-moon system
- what causes lunar and solar eclipses
- how Earth and the moon are alike and different

VOCABULARY

revolve
orbit
rotate
axis
eclipse

Earth and the Moon in Space

The moon is the brightest object in the night sky and Earth's nearest neighbor in space. Together, Earth and the moon are part of the sun's planetary system. Pulled by the sun's gravity, the Earth-moon system **revolves**, or travels in a closed path, around

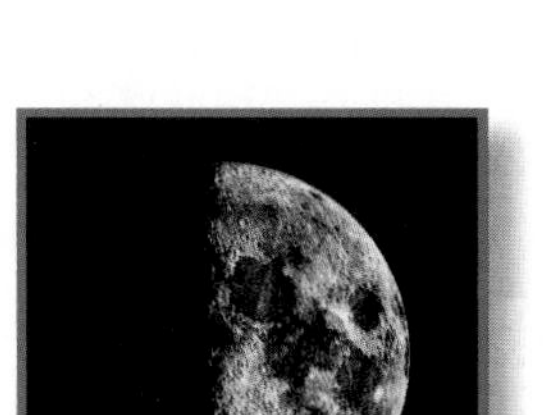

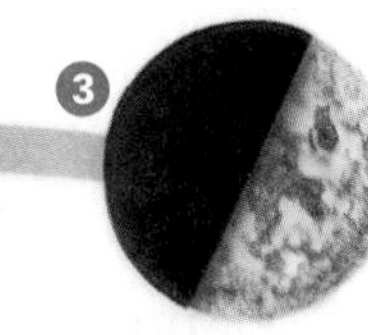

❸ **FIRST QUARTER**
About one week after a new moon, the moon looks like a half-circle. This phase is called the first quarter because the moon is a quarter of the way around Earth.

❹ **WAXING GIBBOUS**
The word *gibbous* (GIB•uhs) comes from a word meaning "hump."

❺ **FULL MOON**
About two weeks after a new moon, we see the entire sunlit half.

❻ **WANING GIBBOUS**
A waning moon appears to get smaller.

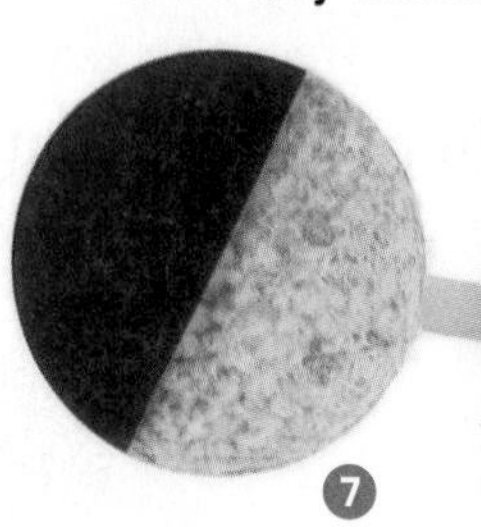

❼ **LAST QUARTER**
About three weeks after a new moon, the moon is three-fourths of the way around Earth.

the sun. The path Earth takes as it revolves is called its **orbit**. Earth's orbit is an ellipse, a shape that is not quite circular.

As Earth orbits the sun, it **rotates**, or spins on its axis. The **axis** is an imaginary line that passes through Earth's center and its North and South Poles. Earth's rotation results in day and night. When a location on Earth faces the sun, it is day in that place. When that location faces away from the sun, it is night.

Pulled by Earth's gravity, the moon revolves around Earth in an ellipse-shaped orbit. When the moon is closest to Earth, it is about 356,400 km (221,463 mi) away.

Like Earth, the moon rotates on its axis. However, the moon takes 27.3 Earth days to complete one rotation. This makes a cycle on the moon of one day and one night that is 27.3 Earth days long.

Even though the moon rotates, the same side of the moon always faces Earth. This is because the moon orbits Earth in 27.3 days—exactly the same amount of time it takes to rotate once on its axis.

Although the moon shines brightly at night, it does not give off its own light. We see the moon from Earth because sunlight is reflected off its surface. As the moon orbits Earth, its position in the sky changes. This produces the different shapes, or phases, of the moon we see each month. The phases of the moon, as seen from Earth, are shown in the photographs below and on page D6.

✔ **How do Earth and the moon move through space?**

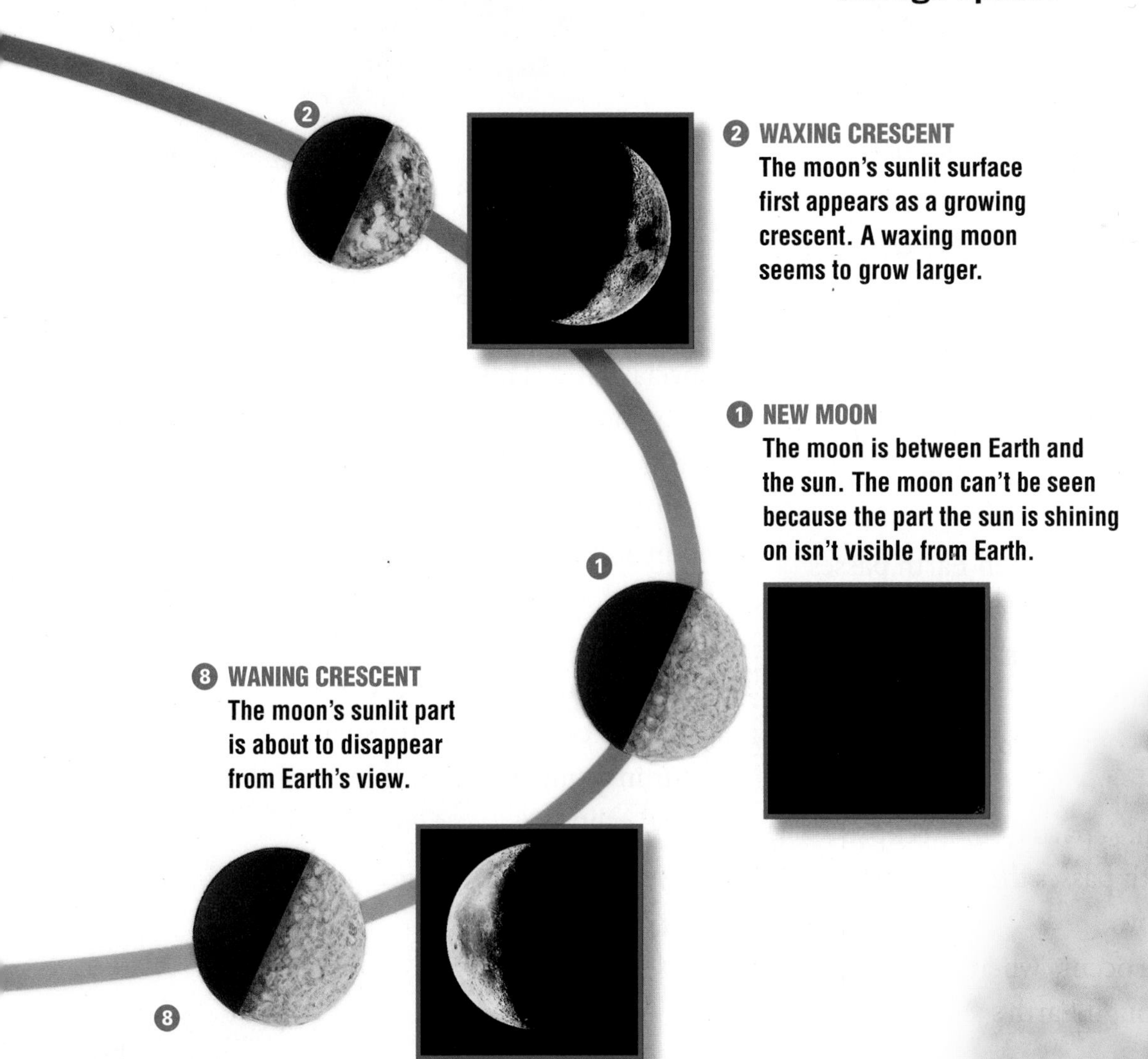

2 WAXING CRESCENT
The moon's sunlit surface first appears as a growing crescent. A waxing moon seems to grow larger.

1 NEW MOON
The moon is between Earth and the sun. The moon can't be seen because the part the sun is shining on isn't visible from Earth.

8 WANING CRESCENT
The moon's sunlit part is about to disappear from Earth's view.

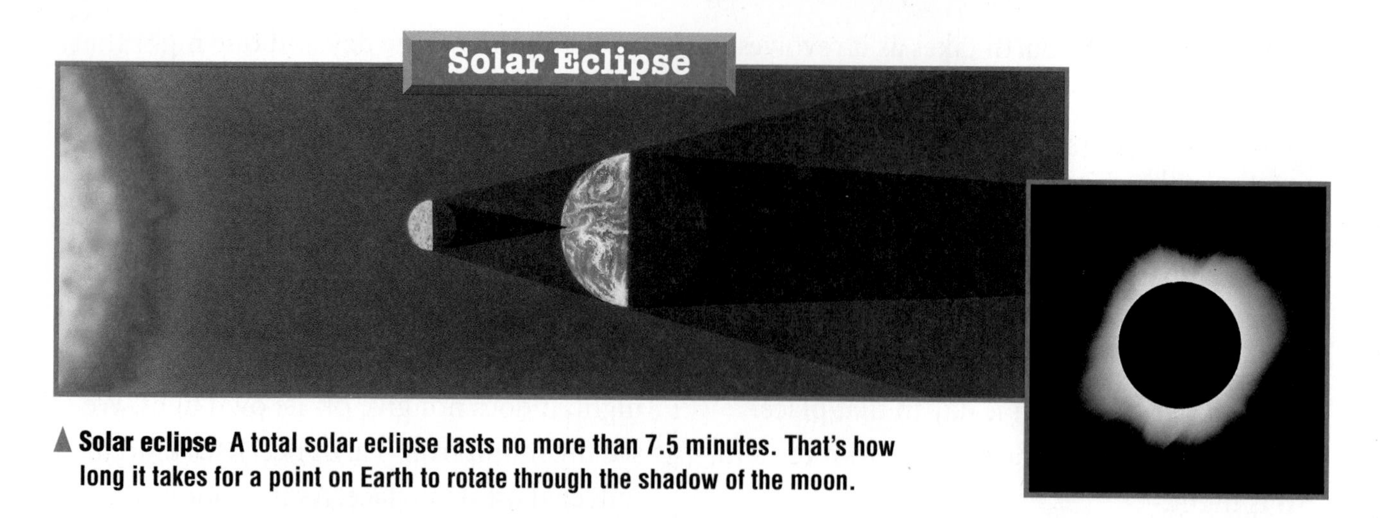

▲ **Solar eclipse** A total solar eclipse lasts no more than 7.5 minutes. That's how long it takes for a point on Earth to rotate through the shadow of the moon.

▲ **Lunar eclipse** A total lunar eclipse lasts more than two hours. It may be seen from any place on Earth that is facing the moon.

Solar and Lunar Eclipses

All bodies in the solar system produce shadows in space. An **eclipse** (ee•KLIPS) occurs when one object passes through the shadow of another. A solar eclipse or a lunar eclipse occurs when Earth, the sun, and the moon line up.

A solar eclipse occurs when Earth passes through a new moon's shadow. During a total solar eclipse, the moon appears to completely cover the sun. The sky darkens, and only the sun's outer atmosphere is visible. It can be seen glowing as a bright circle around the moon. A partial solar eclipse occurs when Earth passes through part of the moon's shadow.

A lunar eclipse occurs when the full moon passes through Earth's shadow. When Earth passes between the sun and the moon, it blocks the sun's light. However, Earth's atmosphere bends certain colors of light, especially red. This makes the eclipsed moon look like a dim red circle.

You may wonder why eclipses do not occur twice each month—at every new moon and full moon. This is because the moon's shadow usually passes above or below Earth, or the moon passes above or below Earth's shadow. Only seven eclipses—two lunar eclipses and five solar eclipses—occur in a single year. And most of those are partial eclipses.

✓ **How does a solar eclipse differ from a lunar eclipse?**

The Moon's Surface

When the moon first formed, its surface was hot, molten rock. As the surface cooled, it formed a rocky crust. If you look at the moon through a telescope, you can see three types of landforms—craters, highlands, and dark, flat areas.

Some moon craters are very large. Tycho (TY•koh) crater, for example, is 87 km (about 54 mi) across. Other craters are so small that a hundred of them could fit on your fingernail.

The moon also has dark, flat areas known as *maria* (MAH•ree•uh). The word *maria* is Latin for "seas." For many years people thought these flat, dark areas on the moon were seas filled with water. But maria are really areas of hardened lava.

Maria formed when hot, molten rock flowed from the interior through cracks in the moon's surface. This molten rock overflowed some craters and spread across the moon's surface. It then cooled to form a dark rock called *basalt*. The largest mare (MAH•ray) is 1248 km (about 775 mi) across. The illustration below shows craters and other landforms on the moon's surface.

✓ **What are some landforms found on the moon's surface?**

Lunar landforms include craters, maria, ray craters, rilles, highlands, and volcanic domes. Ray craters are thought to be new craters. The rays formed from rock that "splashed" out of the crater due to the impact of the object that made the crater. Rilles are lunar valleys. Some lunar highlands are as high as Earth's mountains. Volcanic domes may be like some volcanoes on Earth. ▶

Comparing Earth's and the Moon's Features

Earth and its moon are alike in several ways. Both are rocky and fairly dense. The same materials that make up Earth—calcium, aluminum, oxygen, silicon, and iron—are found on the moon. Craters occur on both, although there are many more craters on the moon than on Earth.

There are also important differences between Earth and the moon. Unlike Earth, the moon has no atmosphere and no liquid water. Because of this, the moon's landscape has not been eroded by wind and water. The moon's surface weathers very slowly, staying the same for millions of years.

Comparing Earth and the Moon

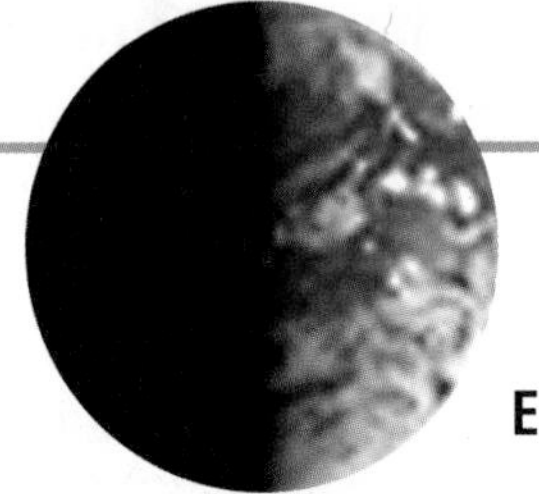

Earth

Craters

Both Earth and the moon have craters. However, the surface of the moon has many more craters than the surface of Earth. Many meteorites burn up in Earth's atmosphere before they reach the surface, and weathering has eliminated many Earth craters.

Weathering

Since the moon has no wind or rain, footprints left by astronauts will remain on its surface for millions of years. Most footprints left on Earth's surface are worn away within a few days.

Water

Life on Earth depends on water. Since the moon has no liquid water, it has no life. However, frozen water has recently been discovered in craters at the moon's poles.

Look at the photographs on page D10 to compare features of the moon and Earth. The photographs on the right were taken on the moon. Those on the left were taken on Earth.

✓ **Compare and contrast features of Earth and the moon.**

Summary

The moon revolves around Earth, while the Earth-moon system orbits the sun. Both Earth and the moon rotate on axes and have day-night cycles. As Earth, the moon, and the sun travel through space, they sometimes line up to produce eclipses. Many of the features on Earth and the moon are different, although some landforms occur on both.

Review

1. Why does the same side of the moon always face Earth?
2. Describe a lunar eclipse.
3. How does a mare differ from a crater?
4. **Critical Thinking** Lunar rocks are very old and do not look as if they were eroded by water. Also, no rocks have been found that contain water combined with minerals, as are found on Earth. What can you **infer** about how long the moon has been without water?
5. **Test Prep** The path that Earth takes around the sun is called its —
   A axis
   B orbit
   C cycle
   D rotation

LINKS

MATH LINK

Use Fractions During a full moon, we see half of the moon's surface. During a first or third quarter, what fraction of the moon's surface do we see?

WRITING LINK

Narrative Writing—Story Some early people were afraid of solar eclipses. They believed that the sun might not return. They made up stories and myths to explain eclipses. For example, a myth might say that a dragon or wolf ate the sun and then spit it out again. Write a story or myth for your class explaining what happens during a solar eclipse.

MUSIC LINK

Moon Music The moon has given people many ideas for songs. Work with a group to list ideas for as many new titles of songs about the moon as you can think of. Then write a song about the moon. You may find it easier to write words to a tune you already know.

TECHNOLOGY LINK

Learn more about the discovery of ice near the moon's poles by viewing *Ice on the Moon* on the **Harcourt Science Newsroom Video.**

LESSON 2

What Else Is in the Solar System?

In this lesson, you can . . .

INVESTIGATE how Earth, the moon, and the sun move through space.

LEARN ABOUT objects in the solar system.

LINK to math, writing, literature, and technology.

The asteroid Ida is unique—it has its own moon!

INVESTIGATE

How Earth, the Moon, and the Sun Move Through Space

Activity Purpose You may not feel as if you're moving right now, but you're actually speeding through space. Earth makes a complete spin once every 24 hours. So if you stand on the equator, you're moving at about 1730 km/hr (1075 mi/hr)! Earth also moves around the sun at about 107,000 km/hr (66,489 mi/hr). At the same time, the moon, which also spins, moves around Earth at about 3700 km/hr (2300 mi/hr). In the investigation you will **make a model** of Earth, the moon, and the sun to **compare** how they move through space.

Materials

- beach ball
- baseball
- Ping Pong ball

Activity Procedure

1. You will work in a group of four to **make a model** of the sun, Earth, and the moon in space. One person should stand in the center of a large open area and hold the beach ball over his or her head. The beach ball stands for the sun. A second person should stand far from the "sun" and hold the baseball overhead. The baseball stands for Earth. The third person should hold the Ping Pong ball near "Earth." The Ping Pong ball stands for the moon. The fourth person should **observe** and **record** what happens.

2. The real Earth moves around the sun in a path like a circle that has been pulled a little. This shape, called an *ellipse* (ee•LIPS), is shown here. For the model, Earth should move around the sun in an ellipse-shaped path. Earth should also spin slowly as it moves around the sun. The observer should **record** this motion. (Picture A)

ellipse

Picture A

3. While Earth spins and moves around the sun, the moon should move around Earth in another ellipse-shaped path. The moon should spin once as it moves around Earth. The same side of the moon should always face Earth. That is, the moon should spin once for each complete path it takes around Earth. The observer should **record** these motions. (Picture B)

Picture B

Draw Conclusions

1. Your model shows three periods of time—a year, a month, and a day. Think about the time it takes Earth to spin once, the moon to move around Earth once, and Earth to move around the sun once. Which period of time does each movement stand for?
2. **Compare** the movements of the moon to the movements of Earth.
3. **Scientists at Work** Scientists often **make models** to show **time and space relationships** in the natural world. However, models can't always show these relationships exactly. How was your model of Earth, the moon, and the sun limited in what it showed?

Investigate Further **Plan and conduct a simple investigation** to test this **hypothesis:** The amount of sunlight reaching Earth changes as Earth moves around the sun.

Process Skill Tip

Making a model of the sun-Earth-moon system enables you to **use time and space relationships** to learn how objects in space move and interact.

Cycles in the Solar System

Rotation and Time

FIND OUT

- about time zones
- what causes seasons
- about planets, asteroids, and comets

VOCABULARY

solstice
equinox
planets
asteroids
comets

You have observed that, each day, the sun appears to rise in the east, reach a high point around noon, and set in the west. This apparent motion of the sun is due to Earth's rotating on its axis. Recall the model you made of Earth and the sun. As Earth rotates, half of it always faces the sun. Locations on Earth's surface facing the sun experience daylight, while locations facing away from the sun experience darkness. Every location goes through a cycle of daylight and darkness in 24 hours. We call this cycle a day.

Our system of telling time is based on Earth's 24-hour rotation. For much of history, people did not need to know the exact time. They got up as the sun rose, ate their midday meal when the sun was overhead, and ended their day as the sun set. In the late 1800s the spread of railway systems produced a need for keeping exact time and schedules. People needed to know when trains would arrive and when they would leave. To solve this problem, 24 standard time zones were set up worldwide. Each time zone represents one of the hours in a day. All places within a particular time zone have the same time. If you travel from one time zone to the next going from east to west, the time will be one hour earlier. If you travel from one time zone to the next going from west to east, the time will be one hour later. The United States has seven time zones, from Puerto Rico in the east, to Hawai'i in the west. When families in Georgia are having dinner, students in Oregon, three time zones to the west, are just getting out of school.

✓ **How are time zones related to Earth's rotation?**

Earth's 360° circumference is divided into 24 time zones. When it is 7:00 A.M. in Atlanta, Georgia, it is only 4:00 A.M. in Portland, Oregon. ▼

Portland

Atlanta

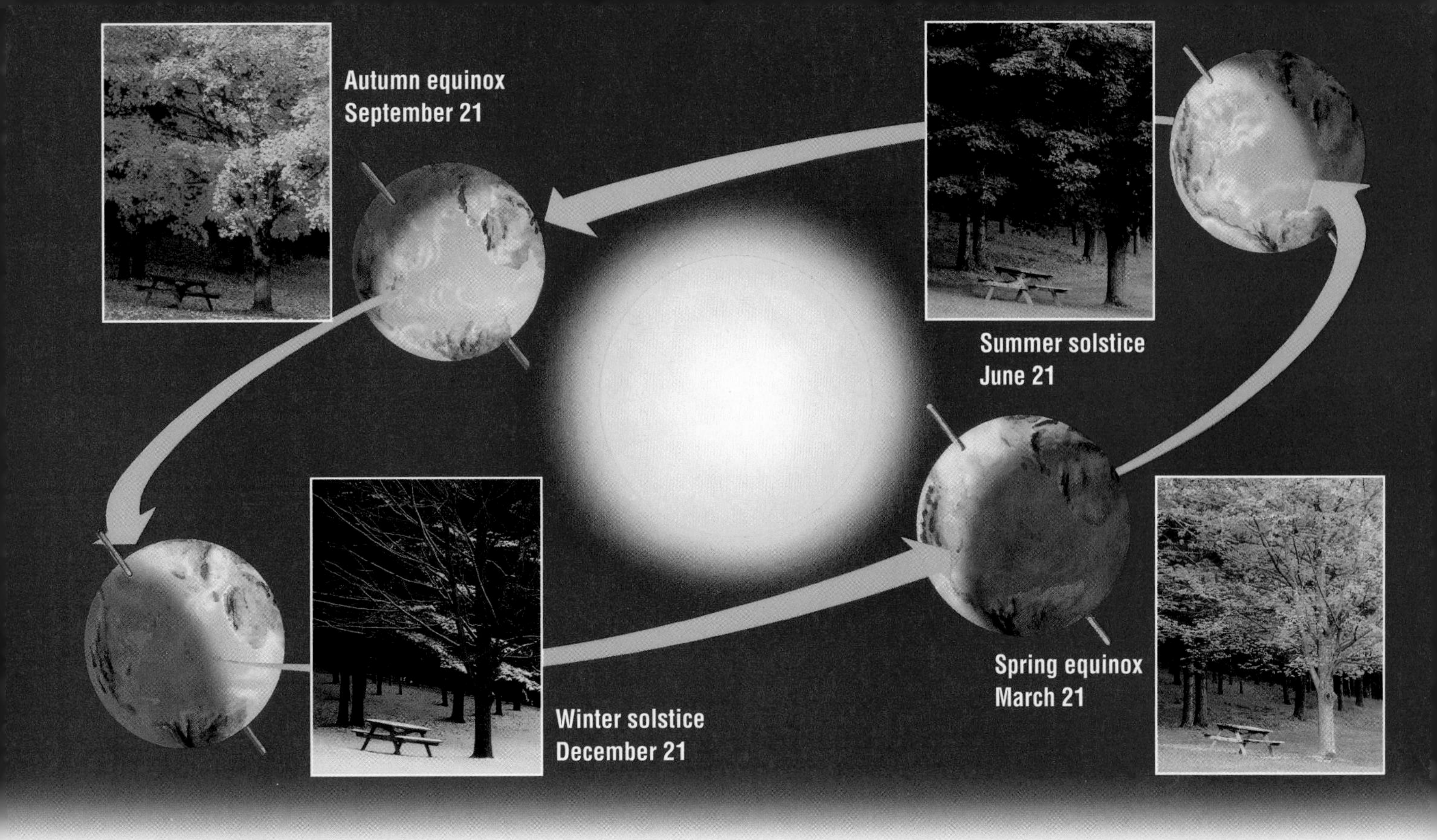

Earth's Seasons

Recall from Lesson 1 that Earth revolves around the sun, following a path called an orbit. It takes about 365 $\frac{1}{4}$ days, or one year, for Earth to complete its orbit. At the same time, Earth rotates once every 24 hours on its axis. However, Earth's axis isn't perpendicular, or straight up and down, in relation to its orbit. It is tilted about 23 $\frac{1}{2}$°. This tilt, along with Earth's changing position in its orbit, causes first the Northern Hemisphere and then the Southern Hemisphere to be pointed toward the sun.

For most places on Earth, this change in position causes changes in the number of hours of daylight and darkness. For example, when the Northern Hemisphere is pointed toward the sun, there are more hours of daylight than darkness, and the sun's rays strike the Earth very directly.

The day with the most daylight in the Northern Hemisphere, about June 21, is the summer solstice. It marks the first day of summer. In the Southern Hemisphere, this day is the winter solstice. Each point in Earth's orbit at which daylight hours are at their greatest or fewest is called a **solstice**. The winter solstice in the Northern Hemisphere is about December 21.

Halfway between the solstice points, neither hemisphere is pointed toward the sun. The hours of daylight and darkness are about equal everywhere. Each point in Earth's orbit at which the hours of daylight and darkness are equal is called an **equinox**. In the Northern Hemisphere, the autumn equinox, about September 21, marks the beginning of fall. The date of the spring equinox in the Northern Hemisphere is about March 21.

✓ **When is the spring equinox in the Northern Hemisphere?**

1 **Sun** The sun contains nearly all the matter in the solar system. Heat, light, and other forms of energy stream outward in all directions from its surface.

2 **Mercury** is so close to the sun that its temperature is about 425°C (800°F). Mercury does not have enough gravity to hold an atmosphere.

3 **Venus** is a hot planet with a thick atmosphere of carbon dioxide. Venus has a surface temperature of about 480°C (900°F), which is much too hot for life.

4 **Earth** With its oxygen-rich atmosphere and liquid surface water, Earth may be the only planet in the solar system able to support life.

5 **Mars** appears red because of the iron oxide, or rust, in its soil. Like Earth, Mars has frozen ice caps at its poles and deserts.

11

10

6 **Asteroid Belt** Asteroids are pieces of rock, perhaps left over from the formation of planets.

7 **Jupiter** is the largest planet in the solar system. Unlike the inner planets, Jupiter is a giant ball of liquid hydrogen and helium surrounded by several thin rings.

8 **Saturn** is another gas giant. Saturn has 18 moons and a huge system of rings, which are made up of ice chunks of varying sizes.

9 **Uranus** has more moons than any other planet and ten thin rings. It is tilted so far on its axis that it rotates on its side.

10 **Neptune** also has thin rings. Its color is similar to that of Uranus. One of Neptune's moons, Triton, is the largest in the solar system.

11 **Pluto** is small and icy. Part of Pluto's orbit passes inside that of Neptune, so at that time Neptune is the planet farthest from the sun.

12 **Comets** A comet has a solid, frozen core. As comets near the sun, their cores begin to melt, forming clouds of gas, which energy from the sun pushes into long tails.

Planets, Asteroids, and Comets

Earth and its moon are only two of the many bodies that make up the solar system. Nine planets, 68 moons, more than 50,000 asteroids and comets, and countless bits of rock, dust, and ice orbit the sun.

Planets are large, round bodies that revolve around a star. In our solar system, the four planets closest to the sun, called the *inner planets,* are small and rocky. Of the five *outer planets,* four are huge and made mostly of gases. The ninth planet, Pluto, is small and icy.

Asteroids are chunks of rock that have been described as looking like giant potatoes in space. Some are nearly as large as small planets, up to 1025 km (about 637 mi) across. Others are the size of a basketball. Some scientists hypothesize that asteroids represent matter that failed to form a planet.

Comets, which are balls of ice and rock, circle the sun from two regions beyond the orbit of Pluto. Glowing clouds surround most comets, and tails of gas can often be seen trailing them as their orbits take them near the sun.

Five planets of the solar system—Mercury through Saturn—are visible from Earth without using a telescope. To the unaided eye, these planets look very much like stars. Viewed through a telescope, as shown in the table below, the planets appear disk-shaped and have a steadier glow than stars.

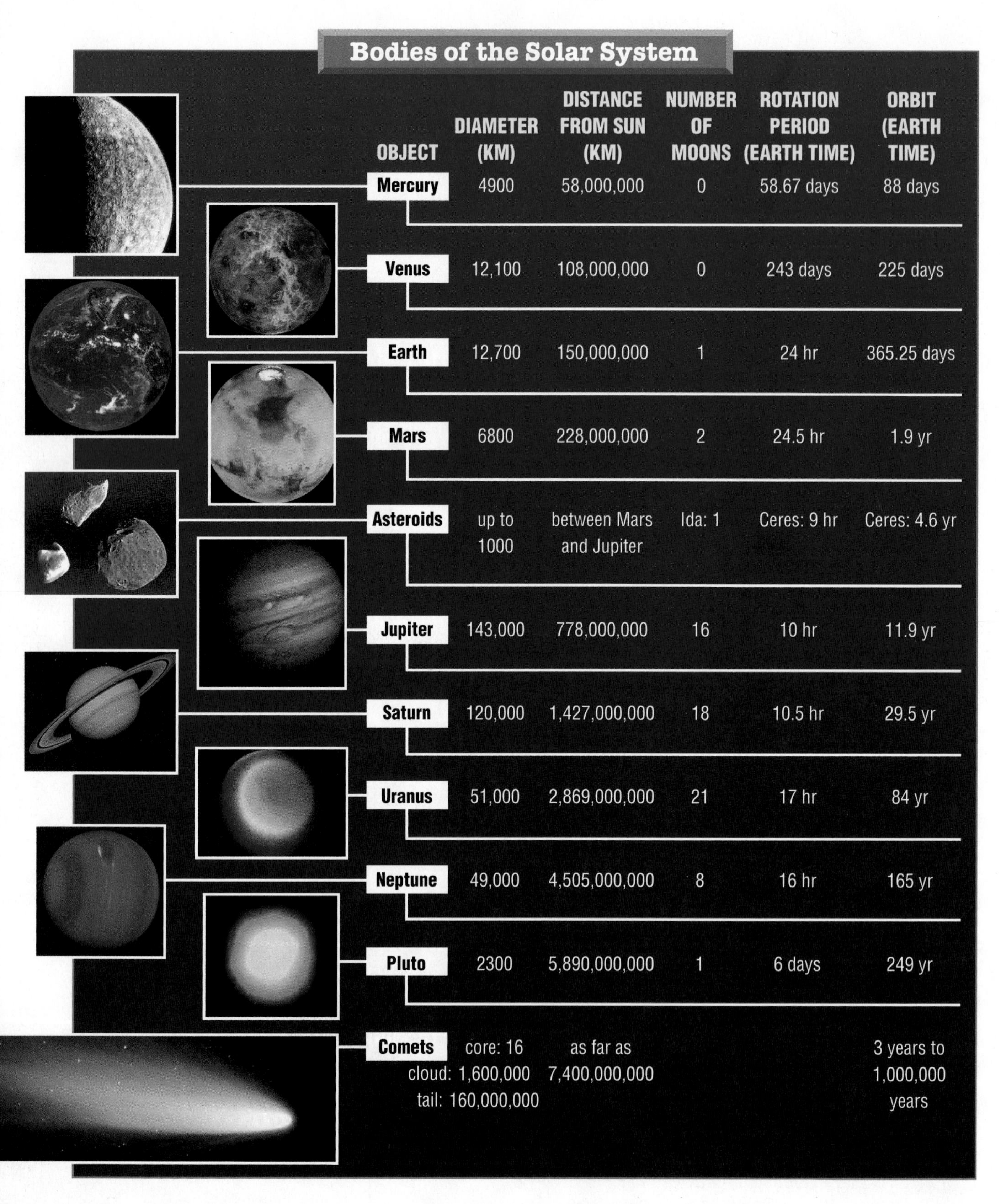

Bodies of the Solar System

OBJECT	DIAMETER (KM)	DISTANCE FROM SUN (KM)	NUMBER OF MOONS	ROTATION PERIOD (EARTH TIME)	ORBIT (EARTH TIME)
Mercury	4900	58,000,000	0	58.67 days	88 days
Venus	12,100	108,000,000	0	243 days	225 days
Earth	12,700	150,000,000	1	24 hr	365.25 days
Mars	6800	228,000,000	2	24.5 hr	1.9 yr
Asteroids	up to 1000	between Mars and Jupiter	Ida: 1	Ceres: 9 hr	Ceres: 4.6 yr
Jupiter	143,000	778,000,000	16	10 hr	11.9 yr
Saturn	120,000	1,427,000,000	18	10.5 hr	29.5 yr
Uranus	51,000	2,869,000,000	21	17 hr	84 yr
Neptune	49,000	4,505,000,000	8	16 hr	165 yr
Pluto	2300	5,890,000,000	1	6 days	249 yr
Comets	core: 16 cloud: 1,600,000 tail: 160,000,000	as far as 7,400,000,000			3 years to 1,000,000 years

moon

Venus, as it appears from Earth without the use of a telescope.

All the planets orbit the sun in the same direction as Earth. All the planets also rotate on their axes, though at very different speeds. The table on page D18 compares some of the important characteristics of the planets and other bodies in the solar system.

✓ **What bodies make up the solar system?**

Summary

Timekeeping on Earth is based on the division of Earth's surface into 24 standard time zones, each representing one of the 24 hours in a day. The position of Earth in its orbit and the tilt of its axis cause the change of seasons. The solar system contains the sun, planets and their moons, asteroids, and comets.

Review

1. What is the difference between a solstice and an equinox?
2. How are asteroids different from planets?
3. What causes a comet to have a tail?
4. **Critical Thinking** Why do astronomers consider Pluto an odd planet?
5. **Test Prep** The gas-giant planets are —
   A Venus, Jupiter, Uranus, Pluto
   B Venus, Jupiter, Saturn, Neptune
   C Jupiter, Saturn, Uranus, Neptune
   D Saturn, Uranus, Neptune, Pluto

LINKS

MATH LINK

Use Divisibility Rules Because distances in the solar system are so large, travel between the planets takes a very long time. If a spacecraft were traveling at 50,000 km/hr, how long would it take it to reach each of the inner planets from Earth? Use the table on page D18 for your information.

WRITING LINK

Informative Writing—Description About 65 million years ago, an asteroid striking Earth may have caused the extinction of the dinosaurs. At least 91 asteroids whose orbits cross Earth's have been identified. Find out how scientists are studying these asteroids and the probability of one of them actually hitting our planet. Then write a news story describing such an event.

LITERATURE LINK

Space Songs by Myra Cohn Livingston is a collection of poems on topics having to do with the sky, including the moon, the planets, and different times of day and weather. Read the poems in *Space Songs*. Then try making up your own space poem.

TECHNOLOGY LINK

Learn more about planets and other objects in the solar system by visiting this Internet site.
www.scilinks.org/harcourt

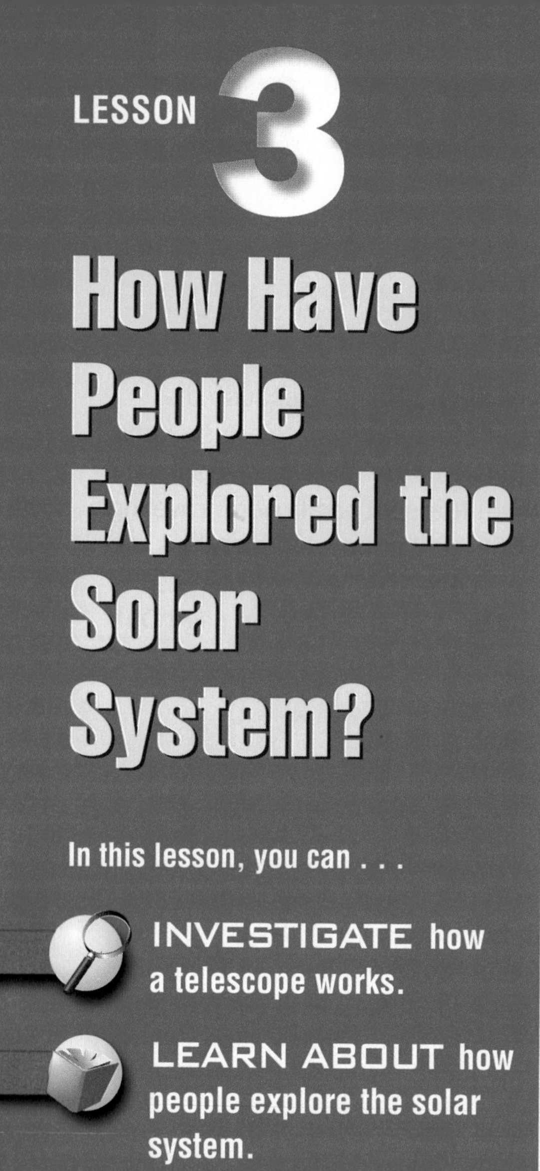

LESSON 3

How Have People Explored the Solar System?

In this lesson, you can . . .

- INVESTIGATE how a telescope works.
- LEARN ABOUT how people explore the solar system.
- LINK to math, writing, art, and technology.

Astronaut working in space ▼

Make Your Own Telescope

Activity Purpose Until Galileo invented the telescope in 1608, people were limited to the power of their eyes for viewing objects at a distance. Since then, people have used telescopes to observe objects in the night sky. To make his telescope, Galileo mounted a curved piece of glass, or lens, at each end of a long tube. In this investigation you will **make a model** telescope and use it to **observe** objects in greater detail.

Materials

- 2 sheets of construction paper
- two convex lenses
- tape
- modeling clay

Activity Procedure

1. Roll and tape a piece of construction paper to form a tube that is slightly larger in diameter than the lenses. Then make a second tube that is just enough larger in diameter for the smaller tube to fit snugly inside it.
2. Slide most of the small tube into the large tube. (Picture A)
3. Place one of the lenses in one end of the smaller tube, and use modeling clay to hold it in place. This lens will be the eyepiece of the telescope. (Picture B)
4. Place the other lens in the far end of the larger tube. Use clay to hold it in place. This lens will be the objective lens, or the lens closest to the object being viewed through the telescope.

5. Choose several distant objects to view with your telescope. You might look at a tree or a distant building. CAUTION **Do not look at the sun with your telescope. You could seriously damage your eyes.** Slide the smaller tube in and out until the object you are viewing comes into focus.

Picture A

6. **Observe** each object twice, first using your eye alone and then using the telescope. **Record** your observations by making two drawings showing how the object appears when viewed with and without the telescope.

7. Repeat steps 5 and 6, observing the moon, a planet, or another object in space, using your telescope. Again make two drawings of the object, showing how it appears both with and without the telescope.

Picture B

Draw Conclusions

1. **Compare** the drawings of each set. How does the appearance of each object change when viewed through the telescope? How does the use of the telescope affect your ability to observe details in those objects?
2. In which of the objects could more details be seen with the telescope than with the eye alone? In which of the objects, if any, were details NOT more visible?
3. **Scientists at Work** Scientists use many kinds of telescopes to **observe** objects in space. Some telescopes use curved mirrors instead of lenses to make objects appear larger. How is your model telescope limited in use for studying objects in space?

Investigate Further **Plan and conduct a simple experiment** to test this **hypothesis:** The curved surfaces of a lens bend the light rays that pass through it.

Process Skill Tip

Making a model of a telescope helps you **observe** the details of distant objects.

Space Exploration

FIND OUT

- about the history of space exploration
- how spacesuits work

VOCABULARY

telescope
satellite
space probe

Exploring the Solar System

Thousands of years ago, people observed the night sky and recorded their observations in cave paintings and rock art. These early observations were made without telescopes or other devices. About the only things early people could see were the phases of the moon and some of the moon's larger features. They could also see some of the planets and many stars. Then, about 400 years ago, the telescope was invented. It allowed people to observe objects in space in much greater detail.

In 1609 Galileo used this telescope to observe the sun, moon, and planets. His telescope had two curved pieces of glass, or lenses, one at each end of a long tube.

900–1200 | 1200–1500 | 1500–1800

The Maya, in Central America, built many *observatories*, or places for viewing the stars and planets. This one at Chichén Itzá, in Mexico, was built about 900.

This telescope was designed by English scientist Sir Isaac Newton in 1668. It used two mirrors and one lens to produce sharper images than Galileo's telescope could.

In 1609 the Italian scientist Galileo (gal•uh•LEE•oh) was possibly the first person to use a new invention—the telescope—to observe the sky. A **telescope** is an instrument that magnifies, or makes larger, distant objects. With this telescope Galileo observed the moon and saw mountains, valleys, and craters that had never been seen before. He also observed the phases of Venus and four moons orbiting Jupiter. About fifty years later, English scientist Sir Isaac Newton used an even better telescope to observe other objects in space.

The modern age of space exploration began in 1957, when the Soviet Union launched *Sputnik I*, an artificial satellite. A **satellite** is any natural body, like the moon, or artificial object that orbits another object. *Sputnik*, which was about twice the size of a soccer ball, carried instruments to measure the density and temperature of Earth's upper atmosphere. The United States launched its own satellite the next year. Soon both countries were launching humans into space.

✓ **How did the telescope help people learn more about objects in space?**

Launched in 1957, *Sputnik I* circled the globe once every 95 minutes for more than a year before it fell back to Earth. The word *sputnik* means "traveling companion" in Russian.

In 1961 the Mercury program sent the first Americans into space.

The first radio telescope, built in 1936, detected radio waves coming from objects in space.

In 1969 the United States landed the first person on the moon.

To the Moon and Beyond

One of the best-known American space programs was Project Apollo. The Apollo missions landed 12 humans on the moon between 1969 and 1972. These astronauts set up experiments and brought back samples of rock. Their work helped scientists learn more about the moon.

In 1977 the *Voyager 1* and *Voyager 2* space probes were launched. A **space probe** is a robot vehicle used to explore deep space. The Voyager space probes have sent back pictures of Jupiter, Saturn, Uranus, and Neptune. Both Voyagers are still traveling through space beyond the solar system.

Other early space probes included *Viking I* and *Viking II,* which landed on Mars in 1976, and the Pioneer probes, which used instruments to "see" through thick clouds that cover Venus. Today's scientists use the Hubble Space Telescope, satellites, and space probes to better understand Earth, the solar system, and what lies beyond.

✓ **What was Project Apollo?**

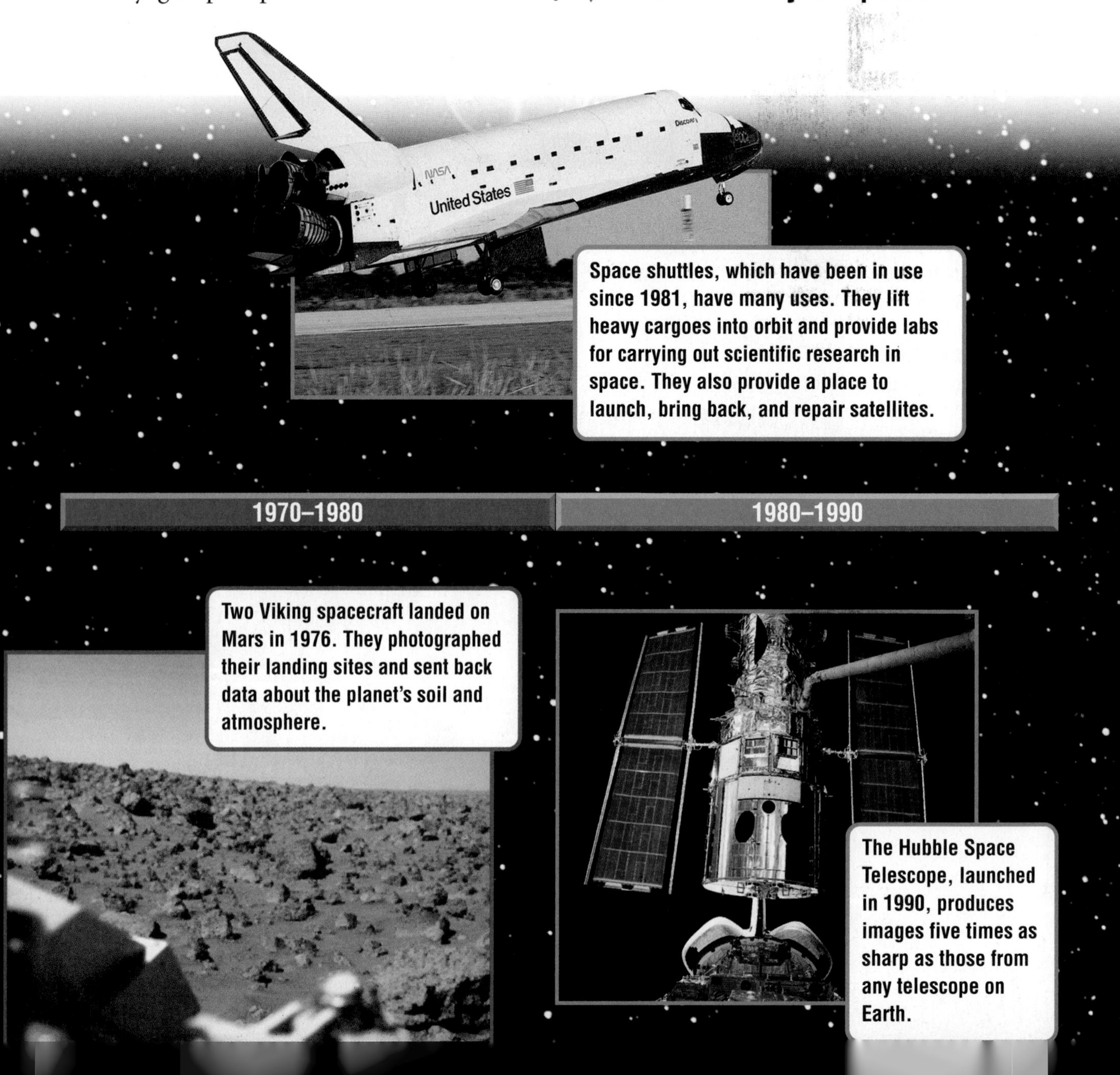

Space shuttles, which have been in use since 1981, have many uses. They lift heavy cargoes into orbit and provide labs for carrying out scientific research in space. They also provide a place to launch, bring back, and repair satellites.

Two Viking spacecraft landed on Mars in 1976. They photographed their landing sites and sent back data about the planet's soil and atmosphere.

The Hubble Space Telescope, launched in 1990, produces images five times as sharp as those from any telescope on Earth.

Spacesuits

The Apollo spacesuit below, once worn by Neil Armstrong, is a $10 million outfit made to protect an astronaut from the moon's hostile environment. The spacesuit must keep an astronaut from "cooking" in direct sunlight or freezing in cold shadows. It must provide the person who wears it with air, water, and waste removal for a moonwalk that may last up to eight hours. The spacesuit must also be flexible enough for an astronaut to walk, twist, turn, bend over, and pick up objects in the reduced gravity on the moon. Flightsuits, such as the one at the right, are much less bulky.

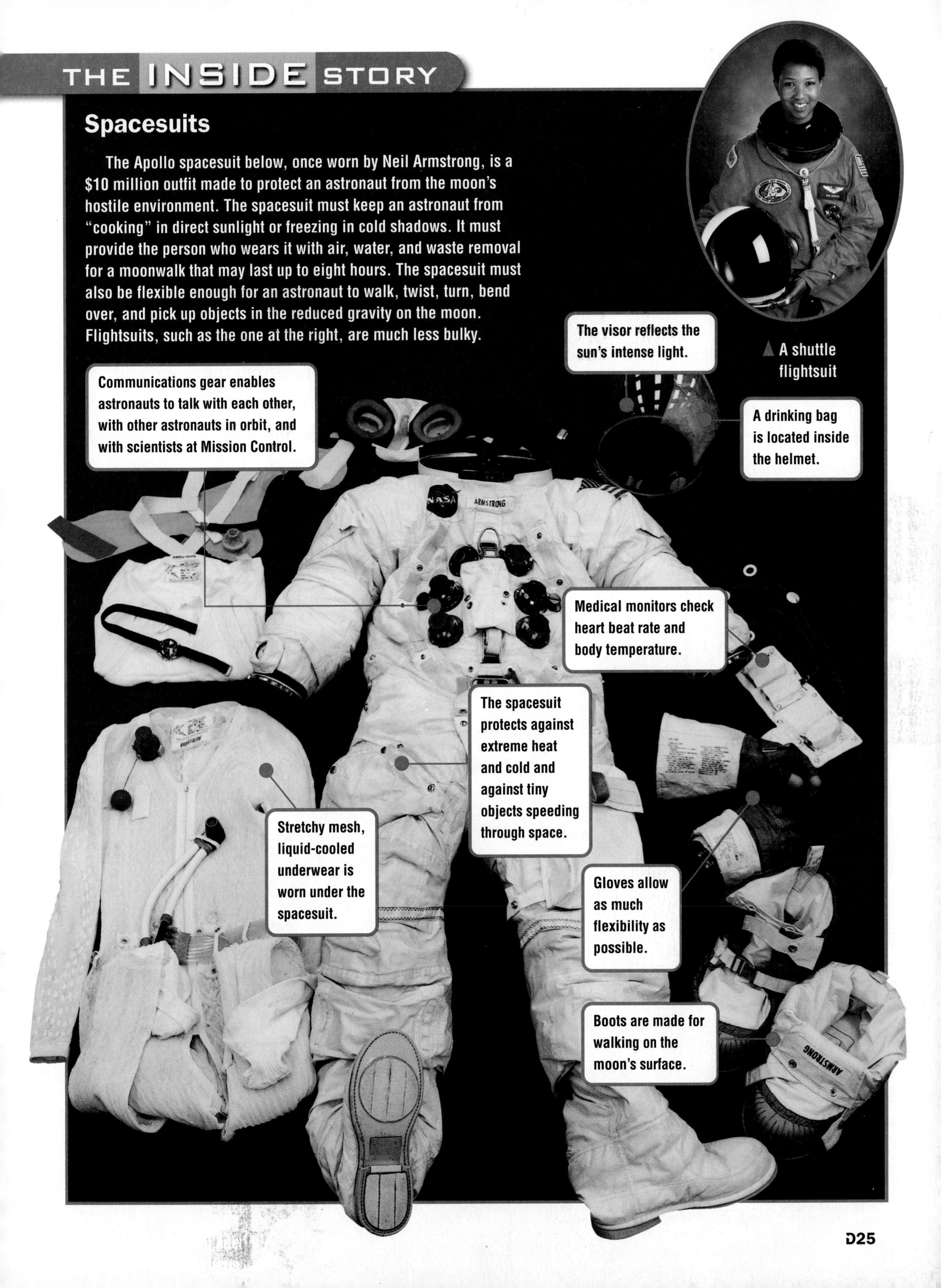

The visor reflects the sun's intense light.
A drinking bag is located inside the helmet.
Communications gear enables astronauts to talk with each other, with other astronauts in orbit, and with scientists at Mission Control.
NASA
ARMSTRONG
Medical monitors check heart beat rate and body temperature.
The spacesuit protects against extreme heat and cold and against tiny objects speeding through space.
Stretchy mesh, liquid-cooled underwear is worn under the spacesuit.
Gloves allow as much flexibility as possible.
Boots are made for walking on the moon's surface.
ARMSTRONG

▲ A shuttle flightsuit

Space Exploration in the Future

The arrival of the first scientists at the International Space Station, *Alpha,* in 2000 marked the beginning of a new era in space exploration. As many as seven scientists at a time will be able to live and work in space. When completed, *Alpha* will be nearly 80 m (about 260 ft) long and have a mass of more than 455,000 kg. In the future, larger stations could have room for a thousand people or more.

Settlements may one day be built on the moon, or even on Mars. Although there are as yet no plans to build bases on the moon, they could be possible by 2020.

A moon base could be used as a research station, like those in Antarctica. To save money, some materials needed to build and run the base could come from the moon itself. For example, some of the moon's rocks contain oxygen. This oxygen could be taken from the rocks and used by people living on the moon. Recently a probe discovered enough ice at the poles to supply a moon base with water. For electricity the base could use solar energy. And some minerals could be mined from the moon and sent back to Earth for processing.

✓ **How could people live on the moon?**

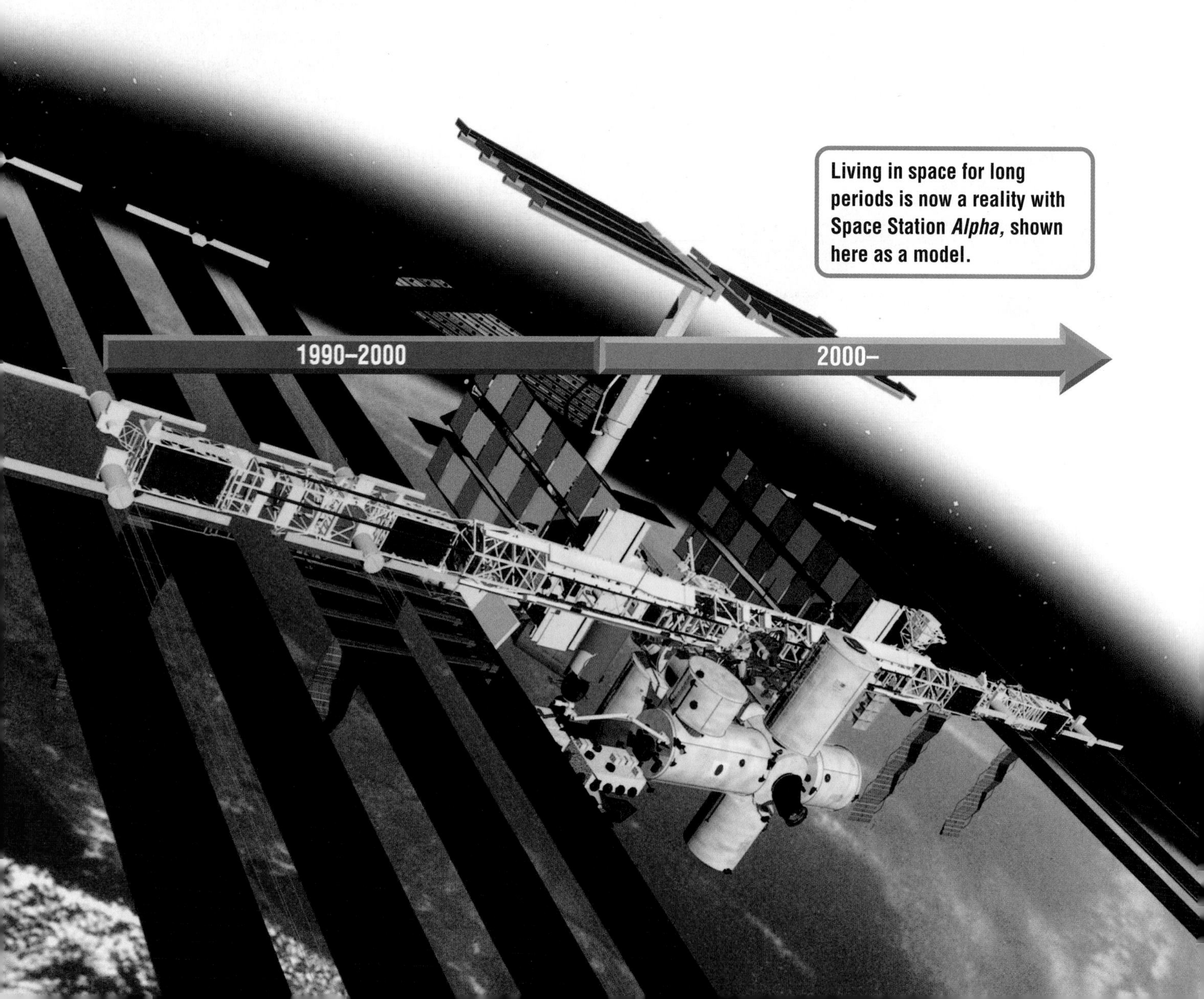

Living in space for long periods is now a reality with Space Station *Alpha,* shown here as a model.

Summary

People have observed and studied the moon and other objects in space since ancient times. The invention of the telescope allowed people to see features and objects that had never been seen before. Today scientists use telescopes, satellites, and space probes to study objects in the solar system and beyond. In the future, people may live and work on space stations and moon bases.

Review

1. What event marked the start of the space age?
2. What are space shuttles used for?
3. What problems have to be solved to build a permanent research station on the moon?
4. **Critical Thinking** A spacesuit weighs more than most astronauts. How can astronauts wear an outfit that is heavier than they are?
5. **Test Prep** The Apollo missions landed humans on —
   A Mercury
   B Venus
   C Mars
   D the moon

LINKS

MATH LINK

Solve Problems Earth spins once every 24 hours. A person standing still on the equator is moving with Earth's rotation at more than 1730 km/hr. How far does he or she move in a 24-hour day?

WRITING LINK

Persuasive Writing—Request Suppose that you have been invited to enter an essay contest. The winner will be the first student astronaut in space. Write a one-page essay to the judges requesting that they choose you. Explain why you are the best candidate.

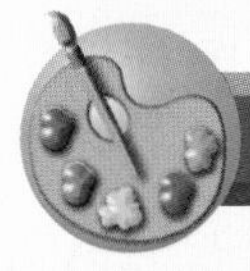

ART LINK

Space Art Design a permanent space station or moon base. Draw a picture to show what it will look like. Label all the major parts, explaining how they will help people live and work in space or on the moon.

TECHNOLOGY LINK

Learn more about the solar system by investigating *Planet Hopping* on the **Harcourt Science Explorations CD-ROM.**

The History of Rockets and Spaceflight

As with many inventions, it is likely that the first rockets were produced partly by accident. Trying to scare off evil forces, the ancient Chinese lit bamboo tubes filled with a combination of charcoal, saltpeter, and sulfur. Perfectly sealed tubes produced loud explosions.

But once in a while an imperfectly sealed tube would shoot off into the air. At some point the Chinese began to produce these tubes, which they called "fire arrows," as weapons.

Rockets of War

Through the centuries, knowledge of how to make rockets spread through Asia, the Middle East, Europe, and the Americas. The "rockets' red glare" that Francis Scott Key wrote about in "The Star-Spangled Banner" were fired by British troops on Fort McHenry near Baltimore, Maryland, during the War of 1812.

The first liquid-fuel rockets were developed during the 1920s by American scientist Robert H. Goddard. These powerful rockets were a great advance in rocket technology. During World War II, a team of German scientists started the Space Age with the launch of the A4 rocket. It traveled 193 km (about 120 mi). Although it was designed to be a weapon of war, the A4 was the first modern rocket, a guided missile. After the war many German rockets were redesigned to collect data from Earth's upper atmosphere.

The Race to Space

In 1957 the Soviet Union used a German-designed rocket to launch *Sputnik*, the first Earth-orbiting satellite. The next year NASA (National Aeronautics and Space Administration) launched the first American satellite, *Explorer I*, and the space race was on. In 1961 a Soviet cosmonaut became the first person in space. Then President Kennedy promised that an American astronaut would be the first person to land on the moon. Competition between the

The History of Rockets and Spaceflight

300 B.C. The Chinese use simple rockets as weapons.

1700s Sir William Congreve develops more powerful rockets for war.

1920s Robert Goddard develops rockets powered by liquid fuel.

1947 The first supersonic (faster than sound) flight is made aboard a rocket-powered airplane, the *X-1*.

1950s The two-stage rocket is developed.

300 B.C. | A.D. 1700 | A.D. 1800 | A.D. 1900

Soviet Union and the United States was fierce. National pride was at stake. In 1964 the United States sent a space probe to Mars. In 1965 a Soviet cosmonaut became the first person to "walk" in space. In 1968 three American astronauts orbited the moon. Finally, in July 1969, the *Eagle* lander of *Apollo 11* touched down on the moon. Neil Armstrong became the first person to walk on another body in space.

Cooperation in Space

The space race ended with the flight of *Apollo 11*. Soon cooperation, not competition, became the key to space exploration. In 1988 Japan, Canada, the United States, Russia, and nine members of the European Space Agency agreed to construct the International Space Station. Many parts have already been built, and the process of launching them on American space shuttles and Russian rockets is underway. Once in orbit the parts are joined by astronauts from several countries.

Rockets began as weapons of war and were improved as sources of national pride. But they have become cargo carriers and transports for scientific research in space.

THINK ABOUT IT

1. Why do you think rockets were first used as weapons?
2. Why do you think the space race ended with *Apollo 11*?

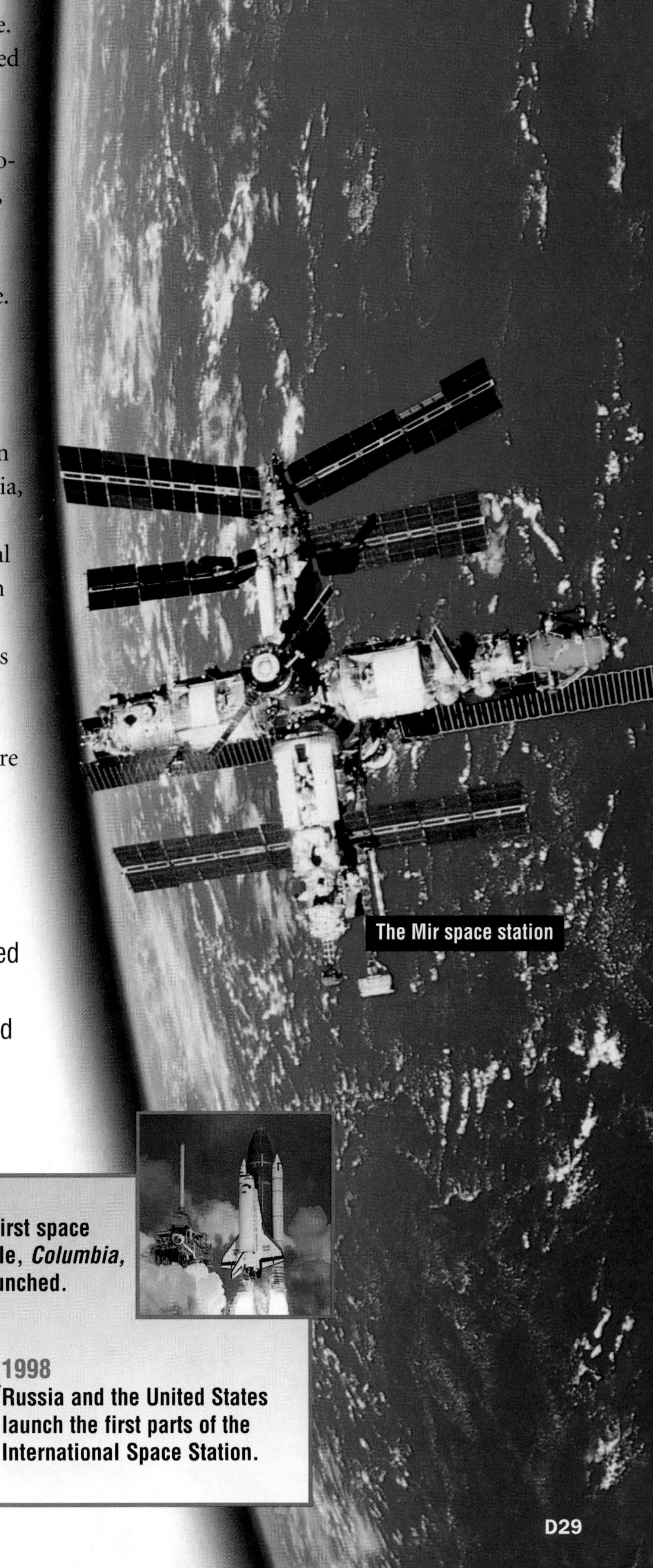

The Mir space station

1962
John Glenn, in a Mercury capsule launched by an Atlas rocket, becomes the first American to orbit Earth.

1969
Neil Armstrong becomes the first person to walk on the moon.

1981
The first space shuttle, *Columbia*, is launched.

1986
The Soviet Union launches the Mir space station.

1998
Russia and the United States launch the first parts of the International Space Station.

A.D. 2000

Harrison Schmitt

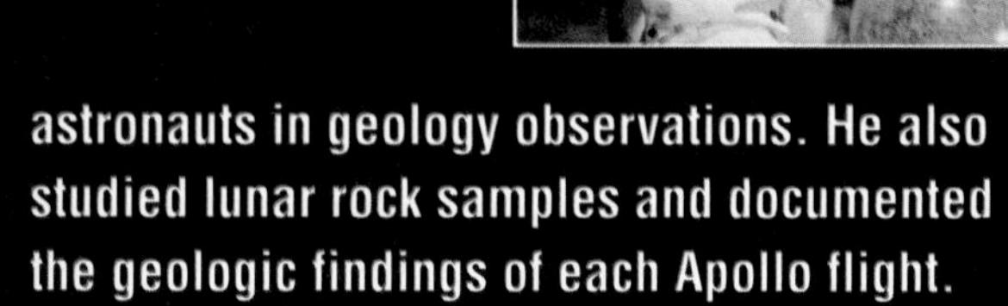

GEOLOGIST, ASTRONAUT

Harrison (Jack) Schmitt is a geologist. He was also the first scientist to fly in space and the only scientist to do research on the moon. In 1971 he was the lunar module pilot for *Apollo 17*. The spacecraft landed in the Taurus-Littrow region of the moon. This area is noted for volcanic cinder cones and steep-walled valleys. At this location Dr. Schmitt collected samples of both young volcanic rock and older mountain rock.

Dr. Schmitt's involvement with the space program began at the United States Geological Survey's Astrogeology Center in Flagstaff, Arizona. There he developed geologic field techniques that were used by all Apollo crews. In 1965 Dr. Schmitt became an astronaut. As the only geologist-astronaut, he trained Apollo astronauts in geology observations. He also studied lunar rock samples and documented the geologic findings of each Apollo flight.

Following his resignation from NASA in 1975, Dr. Schmitt was elected to the United States Senate and served six years as senator from New Mexico. Today Dr. Schmitt is a consultant on issues concerning business, geology, space, and public safety.

THINK ABOUT IT

1. Why was it important for Apollo crew members to have a knowledge of geology?
2. What kinds of information do you think might be gained from samples of lunar rock?

Lunar Rover on the moon

PAPER MOON

How can the moon be used to make a calendar?

Materials

- clock
- 28 white paper plates
- scissors

Procedure

1. Observe the moon at the same time each night for 4 weeks.

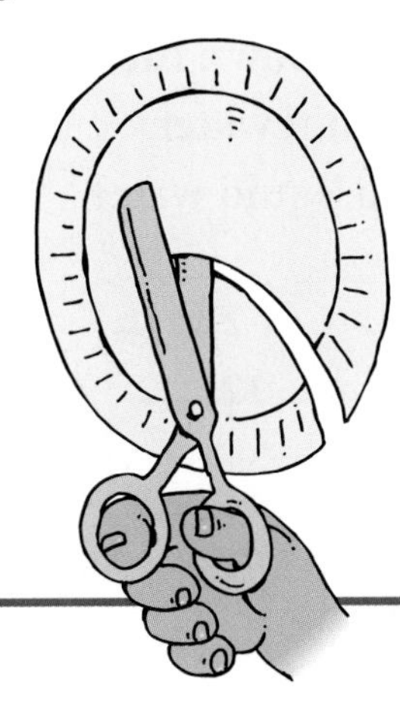

2. Cut one paper plate each night to represent the shape of the moon as you observed it.
3. Hang the paper-plate moons on a wall to make a record of your observations.

Draw Conclusions

How did the shape of the moon change over the length of your observations? What pattern do you notice about the changing shape of the moon? An Earth calendar has 12 months. How many months (moon cycles) would there be in a moon calendar?

SOLAR-SYSTEM DISTANCES

How far is it to Pluto?

Materials

- table of planets' distances from the sun
- roll of toilet paper
- wood dowel
- marker

Procedure

1. Round off all distances on the table to the nearest million kilometers.
2. Use one square of toilet paper to represent the distance from the sun to Mercury.

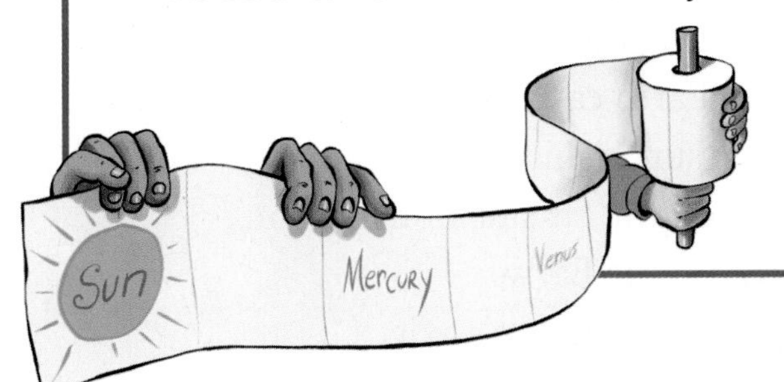

3. Divide the distance from the sun to Mercury into all the other distances. The quotient for each problem will be how many toilet-paper squares each planet is from the sun.
4. Put the dowel into the toilet-paper roll. Unroll the paper, count the squares of paper, and label the position for each planet.

Draw Conclusions

How many squares of toilet paper does it take to show the location of Pluto? How much farther from the sun is Pluto than Mercury? Distances in the solar system are huge. The toilet-paper model helps you visualize those distances. What kind of model could you make to show the sizes of the planets?

CHAPTER 1 Review and Test Preparation

Vocabulary Review

Use the terms below to complete the sentences. The page numbers in ( ) tell you where to look in the chapter if you need help.

revolves (D6)
orbit (D7)
rotate (D7)
axis (D7)
eclipse (D8)
solstice (D15)
equinox (D15)
planets (D17)
asteroids (D17)
comets (D17)
telescope (D23)
satellite (D23)
space probe (D24)

1. Any natural or artificial object that orbits another object is called a ____.
2. Both Earth and the moon have day-night cycles because they each ____, or spin on an ____.
3. The path the moon takes around Earth is its ____.
4. Galileo used a ____ to observe four of Jupiter's moons.
5. A ____ is a vehicle that is used to explore deep space.
6. As a planet travels around the sun, the planet ____.
7. During an ____, one object in space passes through the shadow of another object.
8. In one year, locations in the Northern Hemisphere experience a summer and a winter ____ and a spring and fall ____.
9. Objects in the solar system include nine ____ with their moons, thousands of ____ in orbits between Mars and Jupiter, and many ____, whose orbits take them far beyond Pluto.

Connect Concepts

Write terms and phrases from the Word Bank below where they belong in the Venn diagram.

revolve
rotate
life
no known life
liquid water
no liquid water
weathering
almost no weathering
craters
rocky
atmosphere
no atmosphere

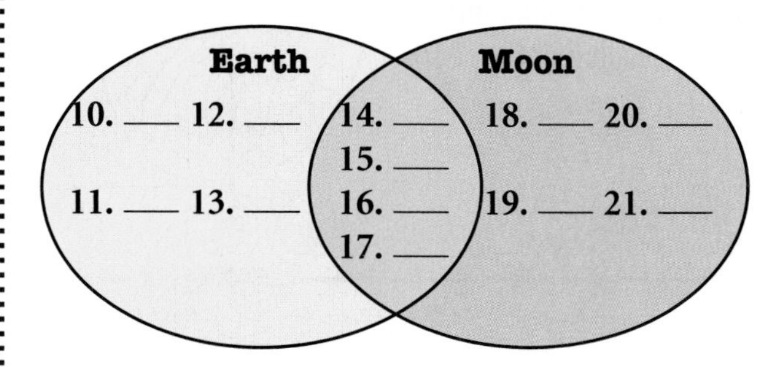

Check Understanding

Write the letter of the best choice.

22. The diagram below shows what season in the Northern Hemisphere?

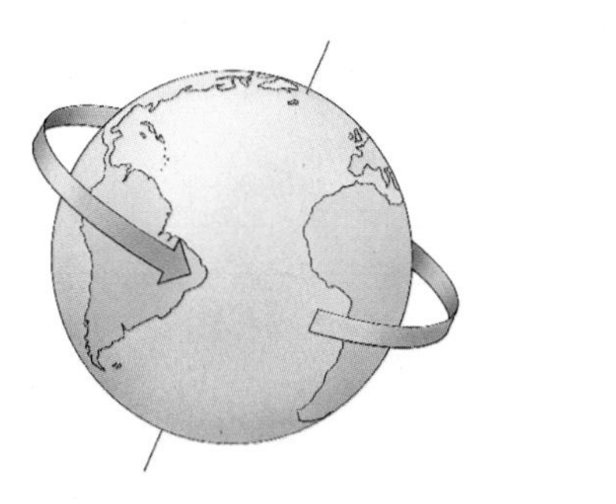

Sun

A summer
B spring
C winter
D autumn

23. During the new-moon phase, a person on Earth cannot see the moon because the sun is shining —
   F on the far side of the moon
   G on the Earth
   H on the moon's axis
   J from behind and below the moon

24. The diagram below shows a —

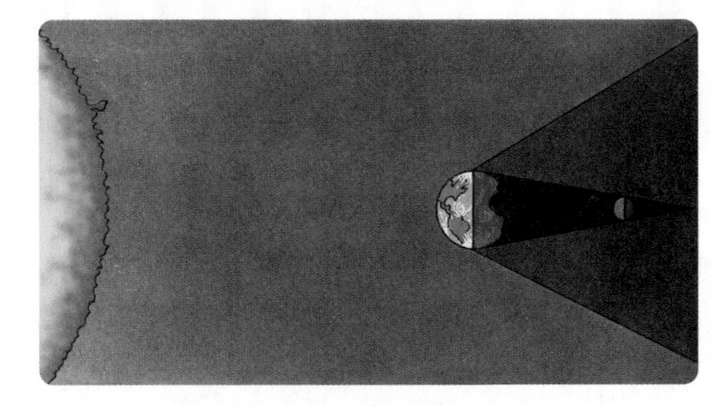

A solar eclipse
B lunar eclipse
C full moon
D new moon

25. The Apollo flights gave scientists first-hand knowledge of —
F the moon
G Earth's atmosphere
H Mars
J the sun

26. Which of the following must a spacesuit provide for an astronaut?
A life support, including air
B protection from intense heat and cold
C a means to orbit the Earth
D both A and B

Critical Thinking

27. Why does the moon appear to wax, or grow larger, and then wane, or get smaller?

28. Mercury has many craters that are millions of years old. Although Earth was hit by large objects from space, just as Mercury was, it has few such craters today. Explain why.

29. On Earth the moon appears to rise and set. If you could look at Earth from the moon, would Earth appear to rise and set? Explain why or why not.

Process Skills Review

30. How can you **use a model** to learn more about the moon?

31. You want to **compare** the moon's landforms with landforms on Earth. What processes will you consider? What tools can you use to **observe** the moon's landforms?

32. On the moon, the sun might rise on July 1 and not set until July 14. **Infer** the effects of such a long day on people living on a moon base.

Performance Assessment

On the Moon

Work with a partner to write a dialogue between an astronaut on the moon and Mission Control on Earth. From the astronaut's point of view, describe the moon's landforms and environment. Include a few details about your spacesuit, too.

CHAPTER

2

The Sun and Other Stars

Vocabulary Preview

photosphere
corona
sunspot
solar flare
solar wind
magnitude
main sequence
universe
galaxy
light-year

Have you ever looked up at night and seen the Milky Way Galaxy? It's hard to miss because it's the *only* thing you can see. All of the stars and planets visible to your eyes—including Earth—are part of it. Beyond the Milky Way Galaxy are hundreds of billions of other galaxies.

Fast Fact

On a clear night you can see more than 2000 stars without using a telescope. However, there are 50,000,000 times as many stars in the Milky Way Galaxy alone.

Fast Fact

In 1989 the *Magellan* spacecraft was launched to Venus from a space shuttle. If you ever get to spend a day on Venus, take plenty of clean clothes. It takes just over 243 Earth days for Venus to rotate once on its axis.

Long Days, Short Days

Planet	Day
Jupiter	9 hr 50 min
Saturn	10 hr 39 min
Neptune	16 hr 3 min
Uranus	17 hr 14 min
Earth	23 hr 56 min
Mars	24 hr 37 min
Pluto	153 hr 18 min
Mercury	1392 hr 30 min
Venus	5832 hr 32 min

Fast Fact

Saturn's rings are made up of chunks of ice, some of which are bigger than most houses. The Voyager space probes showed that Jupiter, Uranus, and Neptune also have rings.

LESSON 1

What Are the Features of the Sun?

In this lesson, you can . . .

INVESTIGATE sunspots.

LEARN ABOUT the sun's structure and features.

LINK to math, writing, and technology.

The surface of the sun

Sunspots

Activity Purpose

The sun always seems the same from Earth. But is it always the same? What changes take place on the sun? You can find out about some of them as you **observe** sunspots in this investigation.

Materials

- white paper
- clipboard
- tape
- small telescope
- large piece of cardboard
- scissors

Activity Procedure

1. CAUTION **Never look directly at the sun. You can cause permanent damage to your eyes.** Fasten the white paper to the clipboard. Tape the edges down to keep the wind from blowing them.
2. Center the eyepiece of the telescope on the cardboard, and trace around the eyepiece.
3. Cut out the circle, and fit the eyepiece into the hole. The cardboard will help block some of the light and make a shadow on the paper.
4. Point the telescope at the sun, and focus the sun's image on the white paper. **Observe** the image of the sun on the paper. (Picture A)
5. On the paper, outline the image of the sun. Shade in any dark spots you see. The dark spots are called *sunspots*. **Record** the date and time on the paper. **Predict** what will happen to the sunspots in the next

15 million °C (27 million °F). At that temperature, and under enormous pressure, particles of hydrogen smash into each other and produce helium. Every time this happens, the sun releases energy as light and heat.

This process is called *fusion* because hydrogen particles fuse, or join together, to produce helium. The fusion of an amount of hydrogen the size of a pinhead releases more energy than the burning of 1000 metric tons of coal. And the sun fuses about 600 million metric tons of hydrogen every second.

Energy from the sun travels in waves, as shown in the illustration below. There are several kinds of waves. Each kind carries a different amount of energy. We see some of the waves as visible light. We feel infrared waves as heat, and ultraviolet waves tan or burn the skin. The sun even produces radio waves, which we hear as radio or TV static. Some of the sun's energy, such as X rays, is harmful to life on Earth. But the atmosphere keeps most of the harmful energy from reaching Earth's surface.

✓ **How does the sun affect life on Earth?**

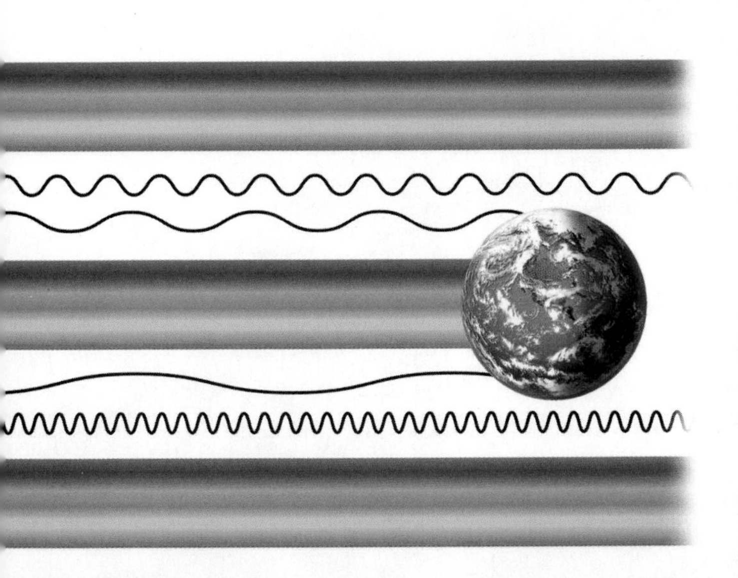

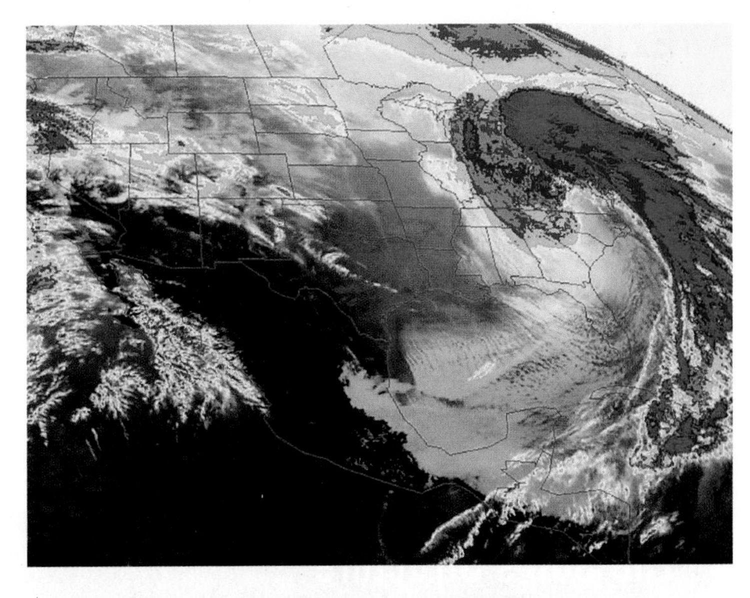

▲ **Solar energy produces differences in air pressure, causing storms, such as hurricanes and blizzards. This photograph is a satellite image of a large winter storm.**

Sometimes particles from the sun stream into space. When these particles reach Earth's atmosphere, they can produce colorful bands of light such as the *aurora borealis,* or northern lights. ►

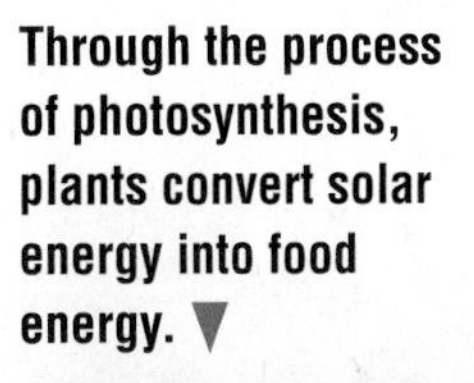

Through the process of photosynthesis, plants convert solar energy into food energy. ▼

Exploring the Sun

The sun's diameter is 1.4 million km (about 870,000 mi)—more than 100 times that of Earth. The sun is large enough to hold 1 million Earths.

Since the sun is so much closer to Earth than other stars are, astronomers study it to understand stars. They have discovered that the sun has several layers of gases. The layers don't have definite boundaries. Instead, each layer blends into the next.

At the center of the sun is the *core*. As you can see in the illustration, the core is small in comparison to the entire sun. However, most of the sun's mass is in its core.

THE INSIDE STORY

The Sun's Structure

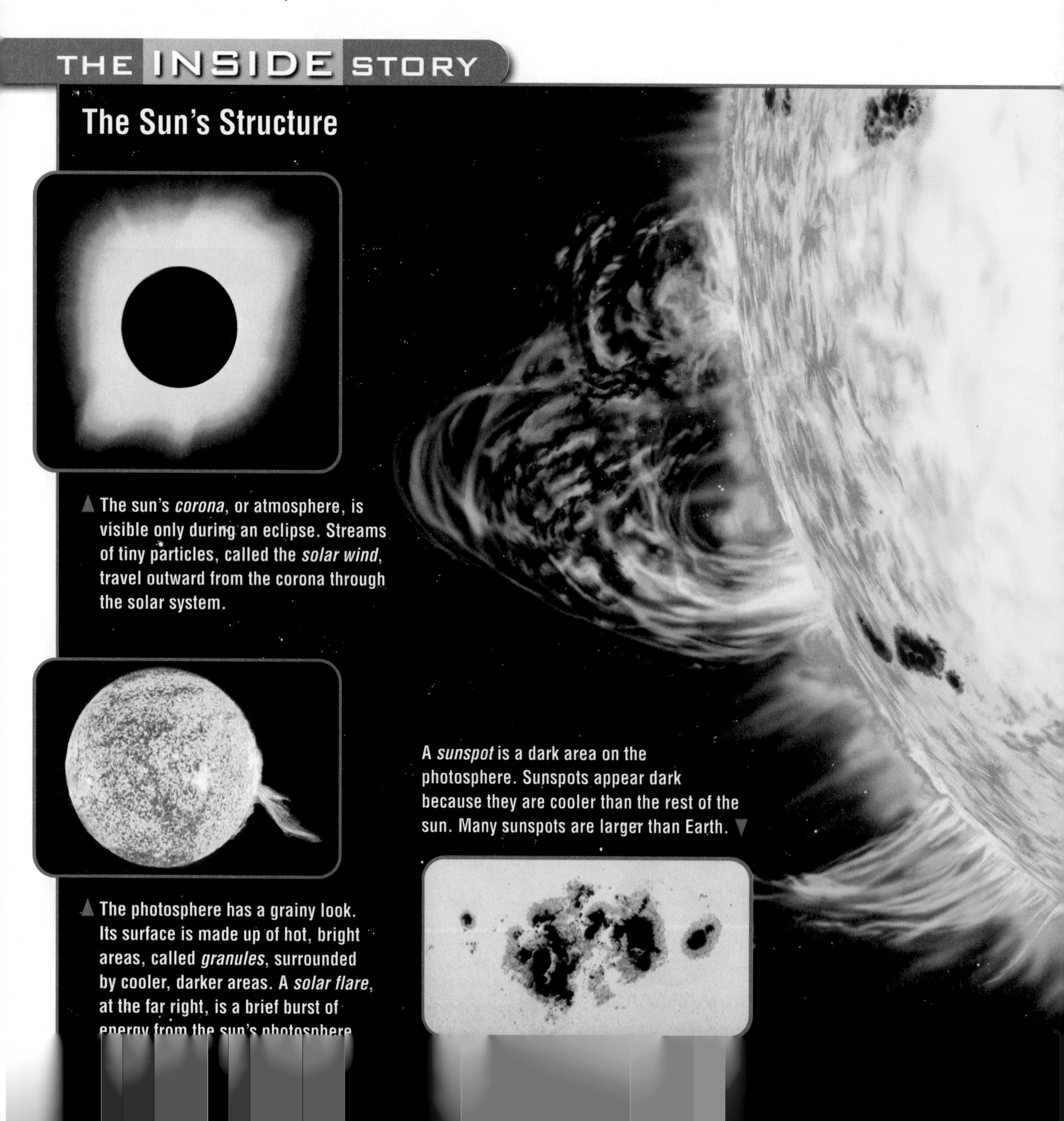

▲ The sun's *corona*, or atmosphere, is visible only during an eclipse. Streams of tiny particles, called the *solar wind*, travel outward from the corona through the solar system.

▲ The photosphere has a grainy look. Its surface is made up of hot, bright areas, called *granules*, surrounded by cooler, darker areas. A *solar flare*, at the far right, is a brief burst of energy from the sun's photosphere.

A *sunspot* is a dark area on the photosphere. Sunspots appear dark because they are cooler than the rest of the sun. Many sunspots are larger than Earth. ▼

As energy from the sun's core moves outward, it passes through the *radiation zone.* Energy from the core heats this layer as a radiator heats the air in a room. From there it moves to the sun's outer layer, the *convection zone.* In the convection zone, energy moves to the surface by a process called convection. In convection, cooler particles are pulled down by gravity, pushing warmer particles up. This is the same way bubbles move energy to the surface of boiling water.

The surface of the sun is known as the **photosphere**, or "sphere of light." This is the surface of the sun we see. Above the photosphere is the sun's atmosphere, the **corona**. This area of hot gases extends 1 million km (about 600,000 mi) from the photosphere.

✓ **What are the layers of the sun?**

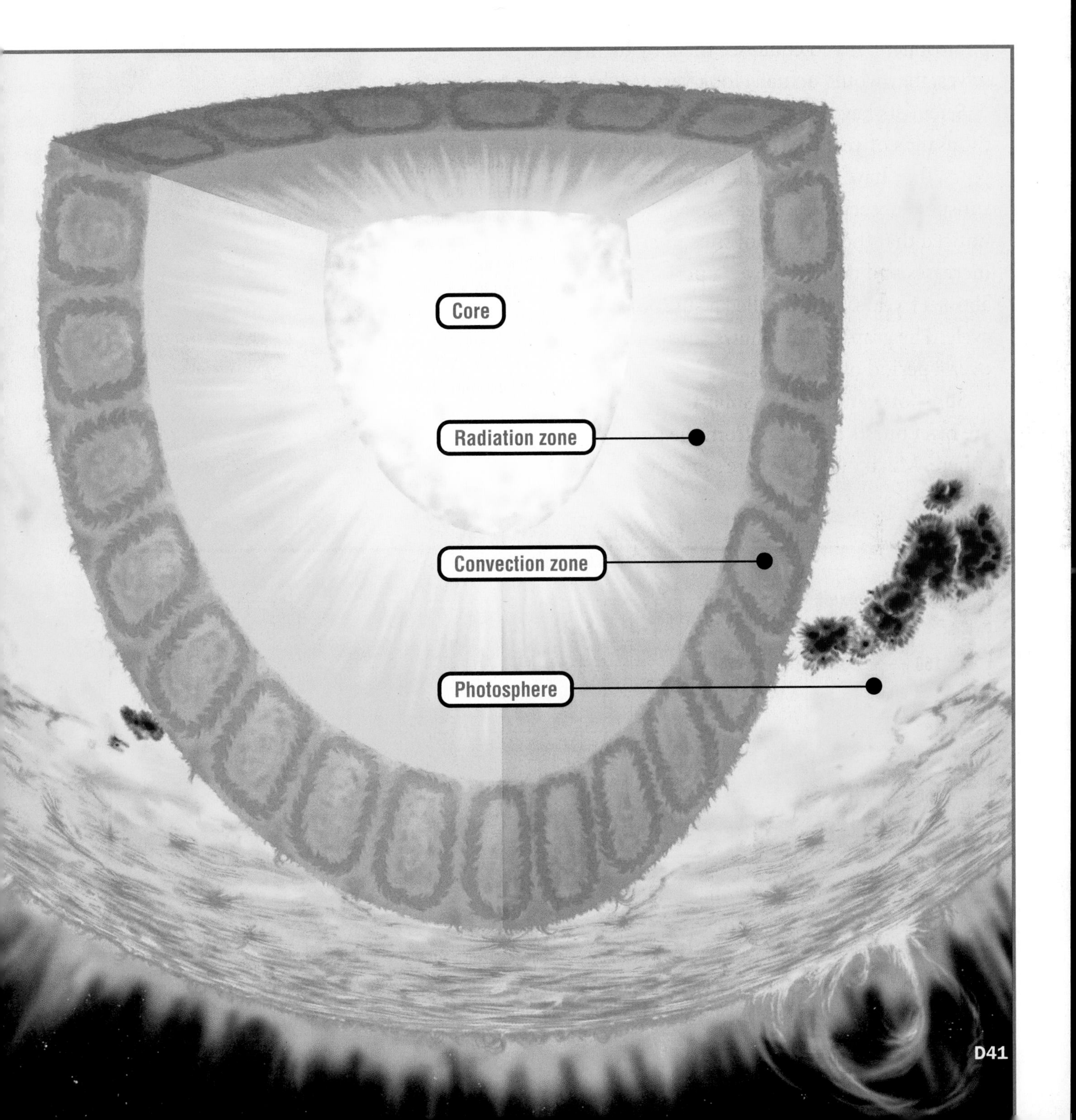

Solar Features

The sun has several features visible at different times at or near its surface. Bright spots on the photosphere are called *granules.* Granules are the tops of columns of rising gases in the convection layer. Darker areas between granules contain cooler gases.

Dark spots, called **sunspots**, are the most obvious features. Sunspots look dark because they are cooler than the rest of the photosphere. If you could see them by themselves, they would actually look very bright.

Scientists have observed sunspots for thousands of years. In the past few hundred years, they have recorded the number of sunspots observed each year. Scientists noticed that the number of sunspots increases and decreases over a period of about 11 years. This is called a sunspot cycle. The graph below shows sunspot cycles over a period of 300 years.

Sunspots can produce **solar flares**. These are brief bursts of energy from the photosphere. Much of a solar flare's energy is ultraviolet waves, radio waves, and X rays. As the energy is released, a fast-moving stream of particles is thrown into space. These particles are called the **solar wind**. When the solar wind reaches Earth, the particles can cause magnetic storms. These storms disturb compasses and energy and communication systems. They also produce auroras in the northern skies.

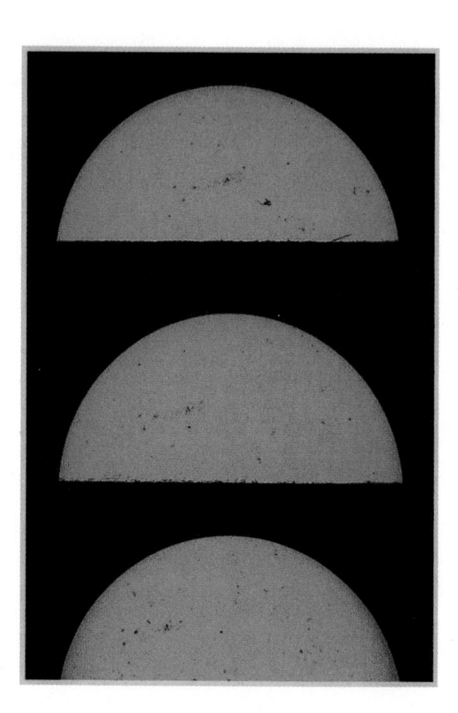

As the sun rotates, groups of sunspots seem to move across its surface. After a few days, the same spots are seen in a different location on the sun's surface, as you saw in the investigation. ▶

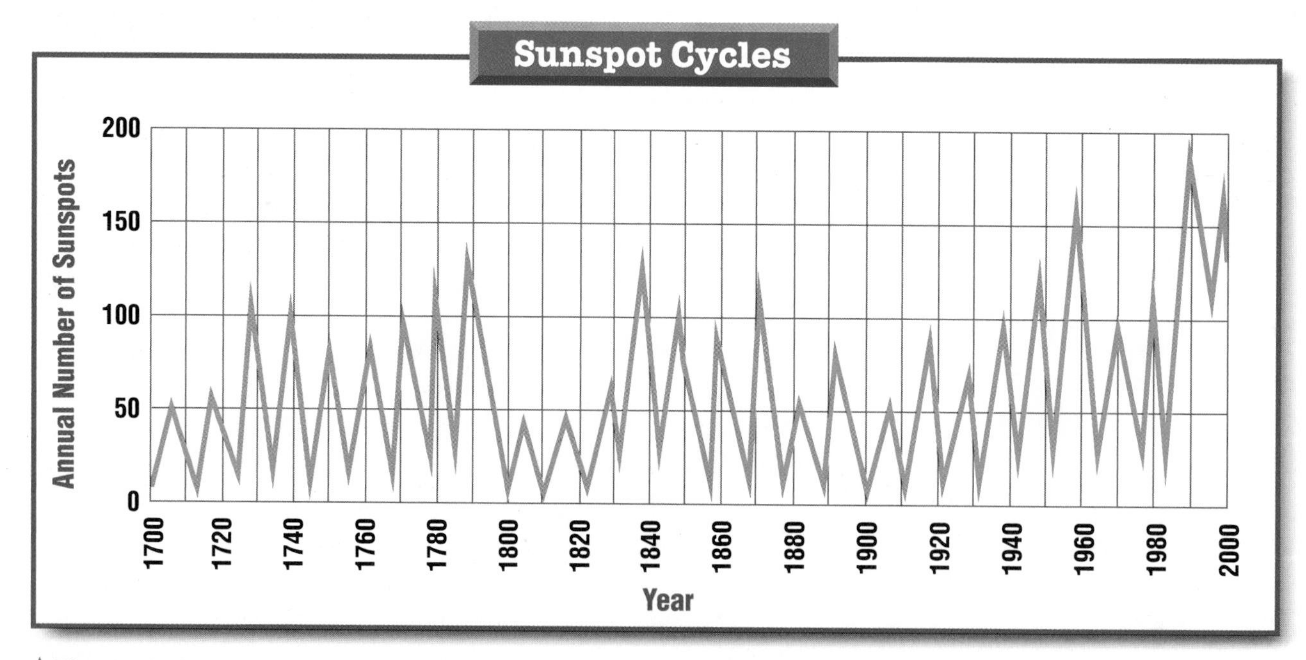

▲ The graph shows how the average number of sunspots varies in cycles. Years with a large number of sunspots are called *sunspot maximums*. Years with a low number of sunspots are called *sunspot minimums*.

A sun feature similar to a solar flare is called a *solar prominence.* A solar prominence is a bright loop or sheet of gas in the corona. It may hover there for days. Or it may explode and disappear in minutes. The photo on page D36 shows a spectacular solar prominence.

✓ **Compare sunspots, solar flares, and solar wind.**

Summary

The sun is a huge mass of hot gases that produces huge amounts of energy. It is the source of most of the energy on Earth. The sun has several layers: the core, the radiation zone, the convection zone, the photosphere, and the corona. Some of the visible features of the sun are solar prominences, solar flares, granules, and sunspots.

Review

1. How does life on Earth depend on the sun?
2. How is energy produced by the sun?
3. Draw a diagram of the sun, showing each layer.
4. **Critical Thinking** Suppose an astronomer observes a huge solar flare. Predict its effect on Earth in the next day or so.
5. **Test Prep** Solar prominences are loops of gas in the sun's —
   A corona
   B radiation zone
   C core
   D photosphere

LINKS

MATH LINK

Multiply Decimals Because distances in the solar system are so large, astronomers use a unit of measure called an astronomical unit, or AU. An AU is the distance between Earth and the sun, about 150 million km. At this distance, energy from the sun reaches Earth in about 8 minutes. Copy the table below, and complete it to show how long it takes the sun's energy to reach each planet.

Planet	Distance (AU)	Time (min)
Mercury	0.4	
Venus	0.7	
Earth	1.0	8
Mars	1.5	
Jupiter	5.2	
Saturn	9.5	
Uranus	19.2	
Neptune	30.0	
Pluto	39.5	

WRITING LINK

Informative Writing—Description Suppose scientists built a space probe that could withstand the sun's high temperatures. Write a description for your teacher of the information the probe might send back as it descends through the sun's layers to its core.

TECHNOLOGY LINK

Learn more about the sun by visiting this Internet site.
www.scilinks.org/harcourt

SCILINKS THE WORLD'S A CLICK AWAY

LESSON 2

How Are Stars Classified?

In this lesson, you can . . .

INVESTIGATE the brightness of stars.

LEARN ABOUT how stars are classified.

LINK to math, writing, technology, and other areas.

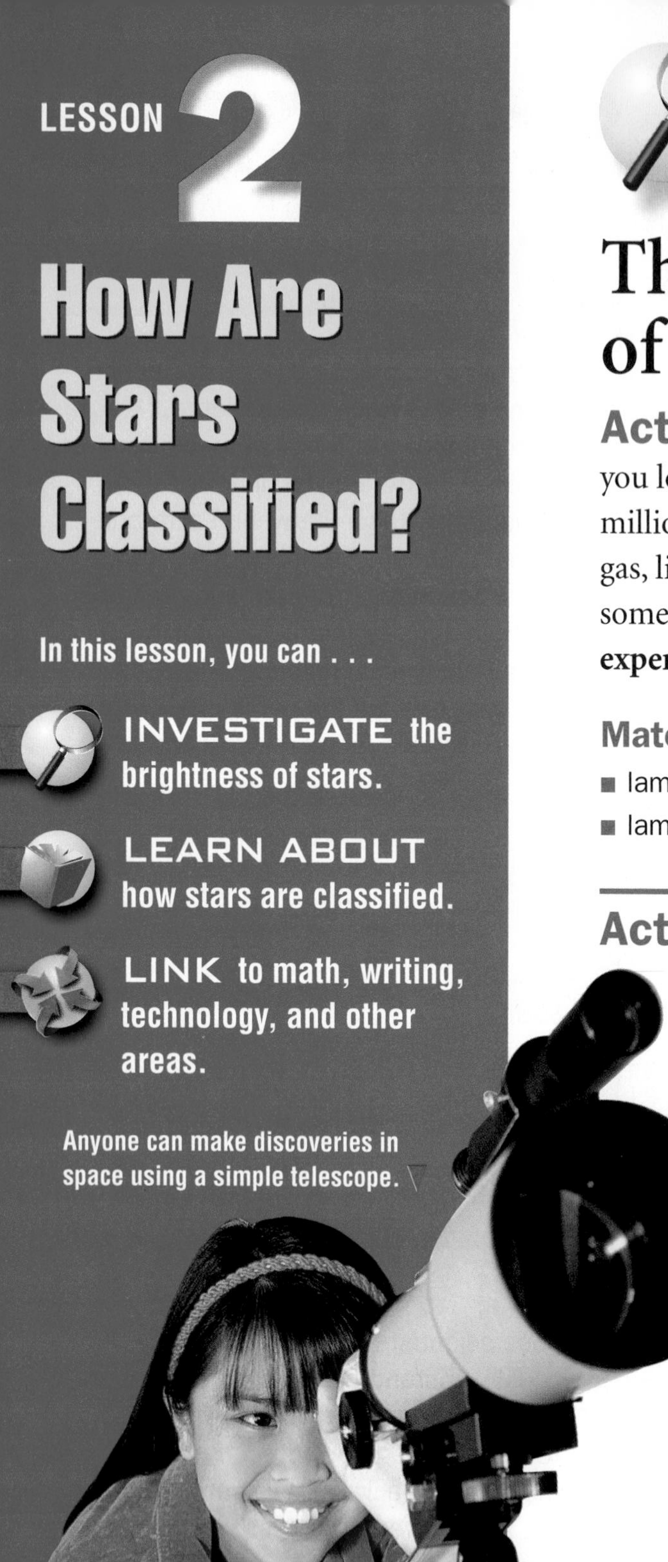

Anyone can make discoveries in space using a simple telescope.

The Brightness of Stars

Activity Purpose Think about the last time you looked at a clear night sky. You probably saw millions of stars. Most stars are hot, bright balls of gas, like the sun. Yet some stars appear bright and some appear dim. Why? In this investigation you can **experiment** to find out.

Materials

- lamp with 40-watt bulb
- lamp with 60-watt bulb

Activity Procedure

1. Place the two lamps near the middle of a darkened hall. Turn the lamps on. (Picture A)
2. **Observe** the lamps from one end of the hall. **Compare** how bright they look. **Record** your observations.
3. Move the lamp with the 60-watt bulb to one end of the hall. **Observe** and **compare** how bright the two lamps look from the other end of the hall. **Record** your observations. (Picture B)

Picture A

Picture B

4. Now place the lamps side by side at one end of the hall. Again **observe** and **compare** how bright the two lamps look from the other end of the hall. **Record** your observations.

5. **Predict** the distances at which the two lamps seem to be equally bright. **Experiment** by placing the lamps at various places in the hall. **Observe** and **compare** how bright the two lamps look from a variety of distances. **Record** your observations.

Draw Conclusions

1. What two variables did you test in this experiment?
2. From what you **observed**, what two factors affect how bright a light appears to an observer?
3. **Scientists at Work** Scientists often **draw conclusions** when **experimenting**. Use the results of your experiment to draw conclusions about how distance and actual brightness affect how bright stars appear to observers on Earth.

Investigate Further Why can't you see stars during the day? **Plan and conduct a simple experiment** to test this **hypothesis:** Stars don't shine during the day.

Process Skill Tip

Drawing conclusions should be based on the results obtained while **experimenting** as well as any other information you have.

How Stars Are Classified

FIND OUT

- about star magnitude
- about the types of stars
- how stars change

VOCABULARY

magnitude
main sequence

Star Magnitude

At night we can see thousands of stars, but during the day we see only one: the sun. The sun is so near that its brightness keeps us from seeing any other stars. As you learned in the investigation, how bright a star looks depends on two factors: its **magnitude**, or brightness, and its distance from Earth.

Suppose two stars that produce the same amount of light are different distances from Earth. They seem to have different magnitudes. The more distant star looks less bright when viewed from Earth, so its *apparent magnitude*, or how bright it seems to be, is less than its *absolute magnitude*, or how bright it really is.

✓ **Why do stars have different apparent magnitudes when viewed from Earth?**

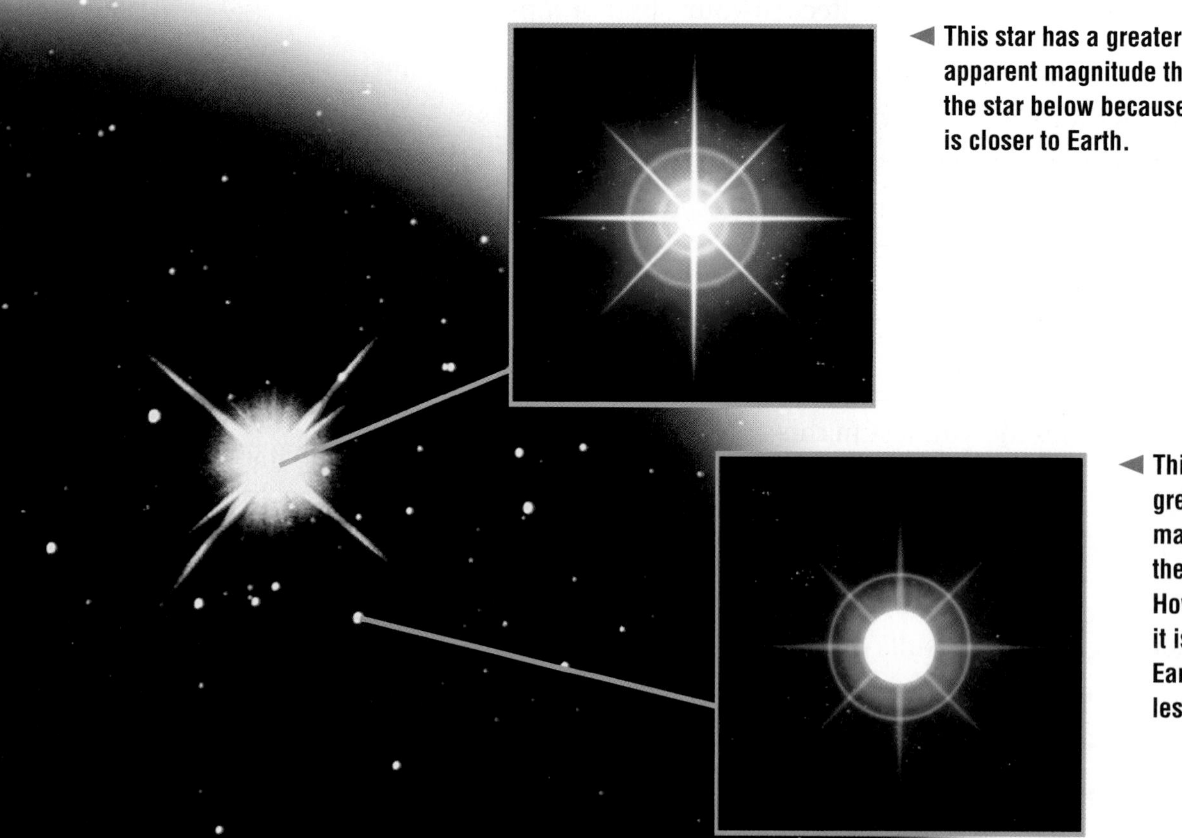

◄ This star has a greater apparent magnitude than the star below because it is closer to Earth.

◄ This star has a greater absolute magnitude than the star above. However, since it is farther from Earth, it looks less bright.

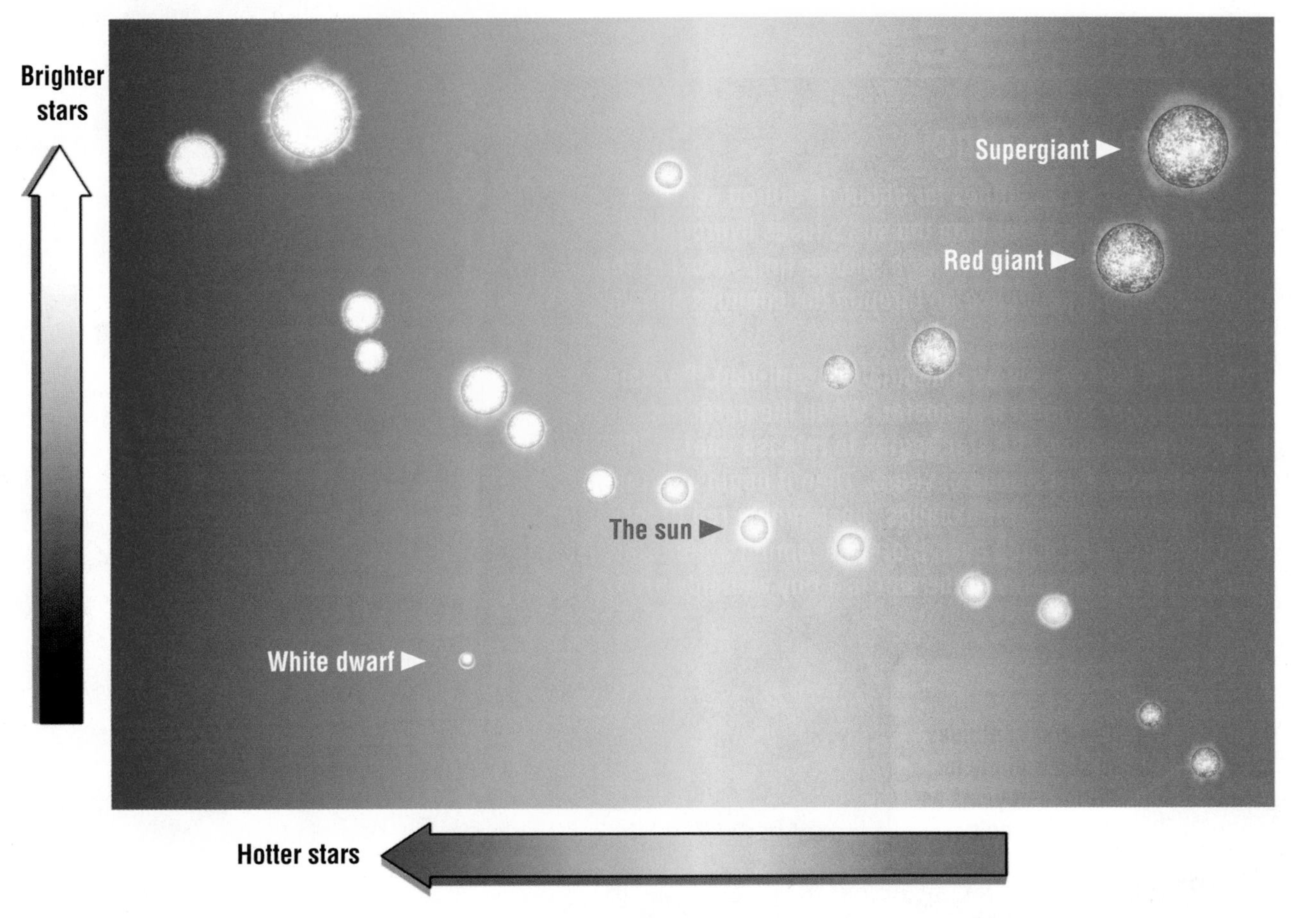

▲ This diagram shows relationships between color, size, magnitude, and temperature of stars.

Types of Stars

On a clear, dark night, you can see that although most stars are white, some are blue and others look red. The color of a star is a clue to its surface temperature. Blue stars have the hottest surface temperatures. Red stars have the coolest. Astronomers use surface temperatures and absolute magnitudes to classify stars.

In the early 1900s, two astronomers—Ejnar Hertzsprung, a Dane, and Henry Russell, an American—classified stars. They used a diagram similar to the one above. The diagram shows the relationships between size, magnitude, temperature, and color.

The absolute magnitudes of stars are plotted from bottom to top. The brightest stars are at the top. The surface temperatures of stars are plotted from right to left. The hottest stars are on the left. The color of a star is shown by its background color. The size of each star is shown by its relative size on the diagram.

Most stars are in a band that runs from the top left of the diagram to the bottom right. This band is called the **main sequence**. Look at the diagram. At the top left are bright, hot, blue stars. In the middle of the main sequence are less bright, cooler stars, such as the sun. These stars shine with a yellow-white light. At the lower right in the main sequence are dim, cool, red stars. About 95 percent of the stars scientists have observed can be classified as main-sequence stars.

✓ **What does a star's position in the main sequence tell you about that star?**

How Stars Change

A star like the sun shines for about 10 billion years. A star with a mass greater than the sun's fuses hydrogen faster. It shines brighter, but for a shorter time. Most of the stars in the main sequence go through the changes shown here.

A typical star begins as a nebula (NEB•yuh•luh). Particles clump together to form a protostar. When the particles become hot enough, the star begins to release energy. It shines steadily for billions of years. Then it begins to run out of hydrogen. The star expands, becoming a red giant. Then its corona expands even more, becoming a planetary nebula. All that remains of the star is a tiny white dwarf.

1 NEBULA
A star begins within a huge cloud of hydrogen, helium, and tiny particles of dust.

7 WHITE DWARF
The star continues to shine dimly for billions of years as it slowly cools.

A red giant is a very old star. When the sun becomes a red giant, it will expand until it consumes both Mercury and Venus. Earth will remain, but the planet will become too hot for life. ▼

6 PLANETARY NEBULA
The red giant's atmosphere expands a million times. It forms a planetary nebula. At the center of the nebula, what is left of the star becomes a white dwarf.

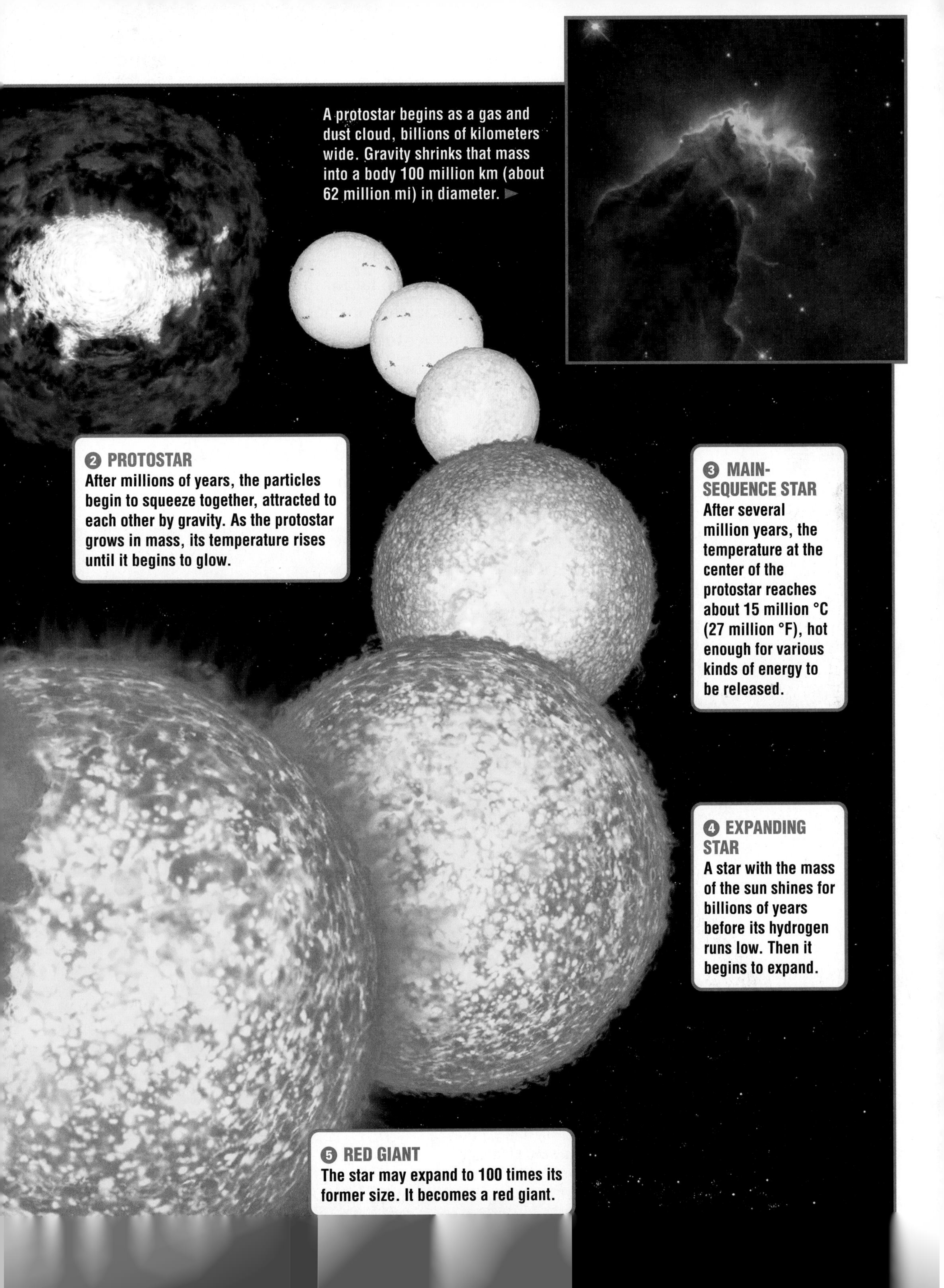
A protostar begins as a gas and dust cloud, billions of kilometers wide. Gravity shrinks that mass into a body 100 million km (about 62 million mi) in diameter.
2 PROTOSTAR
After millions of years, the particles begin to squeeze together, attracted to each other by gravity. As the protostar grows in mass, its temperature rises until it begins to glow.
3 MAIN-SEQUENCE STAR
After several million years, the temperature at the center of the protostar reaches about 15 million °C (27 million °F), hot enough for various kinds of energy to be released.
4 EXPANDING STAR
A star with the mass of the sun shines for billions of years before its hydrogen runs low. Then it begins to expand.
5 RED GIANT
The star may expand to 100 times its former size. It becomes a red giant.

Observing Stars

As you read in Lesson 1, energy from the sun travels in waves. The same is true of the energy of all stars. Scientists learn about stars by studying the energy they send into space. Each kind of wave carries a different amount of energy. The waves range from high-energy X rays to low-energy radio waves. Astronomers use different kinds of telescopes to detect each kind of wave.

◄ **Satellites such as the RXTE are used to detect and measure X rays that stars release into space.**

To detect visible light waves, astronomers need only their eyes. But telescopes can gather much more light than human eyes can.

The larger the telescope, the more light it can gather, and the farther into space it can "see." A large telescope, such as the Keck telescope in Hawai'i, can detect stars 2 million times too far for the human eye to see. The Keck telescope has enough light-gathering power to spot a candle on the moon, and it sees twice as far into space as any other telescope on Earth.

Other kinds of telescopes and instruments are used to gather X rays, radio waves, and other waves of energy.

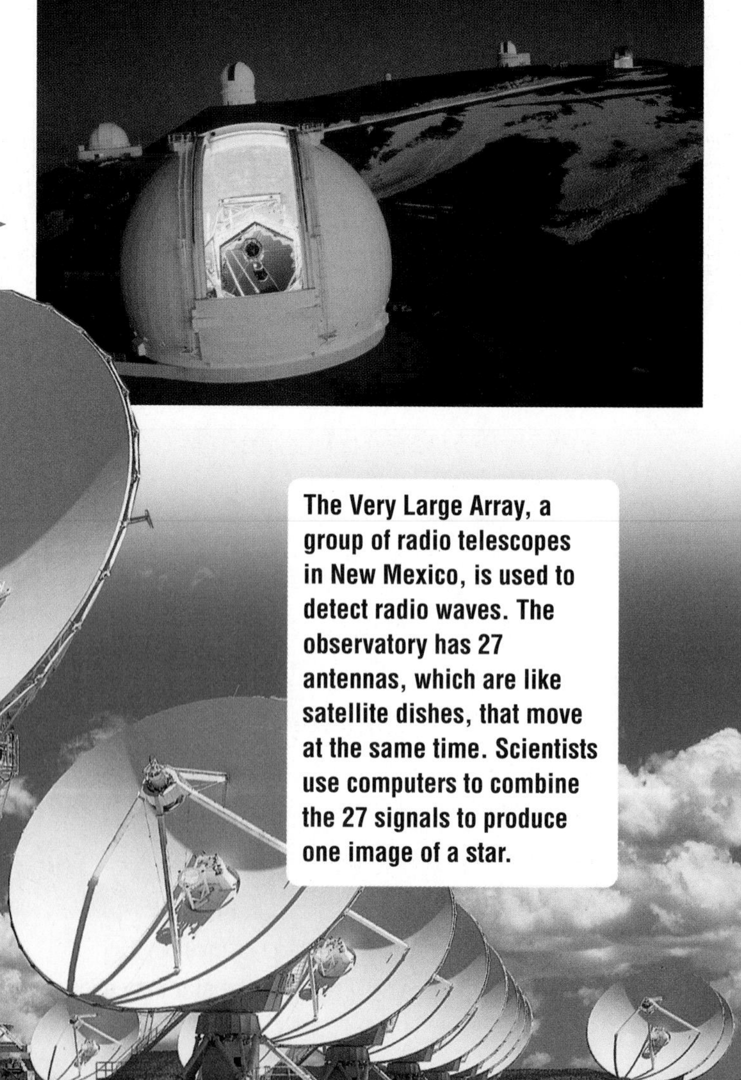

The Keck telescope in Hawai'i is the world's largest reflecting telescope. It has a mirror 10 m (about 33 ft) across for gathering light from distant stars. ►

The Very Large Array, a group of radio telescopes in New Mexico, is used to detect radio waves. The observatory has 27 antennas, which are like satellite dishes, that move at the same time. Scientists use computers to combine the 27 signals to produce one image of a star.

Each wave of energy gives scientists information about the object it came from. For example, the kinds of waves absorbed or given off by a star can tell what it's made of. This is how scientists know that the sun and most other stars are mostly hydrogen and helium.

✓ **How do scientists learn about stars?**

Summary

Scientists classify stars by absolute magnitude, surface temperature, size, and color. Most stars are in the main sequence. Stars undergo a series of changes from nebula to protostar to main-sequence star to expanding star to red giant to planetary nebula and white dwarf. Scientists learn about stars by using telescopes and other instruments to study the energy that stars release into space.

Review

1. What is the difference between absolute magnitude and apparent magnitude?
2. How will the sun's cycle end?
3. Why does a star begin shining?
4. **Critical Thinking** Most of the stars in the main sequence are to the lower right of the sun. What can you conclude about the brightness of these stars and their temperatures?
5. **Test Prep** A star's apparent magnitude depends on —
   A brightness and size
   B absolute magnitude and distance from Earth
   C size and distance from Earth
   D absolute magnitude and size

LINKS

MATH LINK

Multiply Decimals Other than the sun, Proxima Centauri is the star closest to Earth. It is about 4.2 light-years away. A light-year is the distance light travels in one Earth year—about 9.5 trillion km. How far from Earth is Proxima Centauri?

WRITING LINK

Expressive Writing—Poem Write a poem about the stars for your classmates. Use what you have learned in this lesson about stars' magnitudes, temperatures, sizes, and colors.

LANGUAGE ARTS LINK

Star Questions Write a list of five questions you have about stars or other objects in space. If possible, find an astronomer at a local college or an amateur astronomy club. Ask the astronomer your questions.

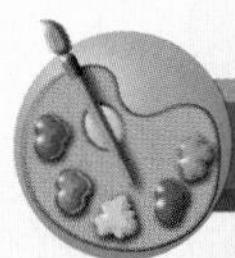

ART LINK

Star Chart Make a diagram or a drawing to show the difference between absolute magnitude and apparent magnitude.

TECHNOLOGY LINK

Learn more about stars by visiting the Harcourt Learning Site.
www.harcourtschool.com

LESSON 3

What Are Galaxies?

In this lesson, you can . . .

INVESTIGATE the location of the sun in the Milky Way Galaxy.

LEARN ABOUT different types of galaxies.

LINK to math, writing, art, and technology.

These star tracks were made with time-lapse photography. ▽

INVESTIGATE

The Sun's Location in the Milky Way Galaxy

Activity Purpose The sun is in a large grouping of stars called the Milky Way Galaxy. But where in this grouping is the sun? In this investigation you can **make a model** of the Milky Way Galaxy. You can **use the model** to help you find out where the sun is in the galaxy.

Materials

- scrap paper

Activity Procedure

1. Make about 70 small balls from scrap paper. These will be your "stars."
2. On a table, **make a model** of the Milky Way Galaxy. Arrange the paper stars in a spiral with six arms. Pile extra stars in the center of the spiral. Use fewer stars along the arms. (Picture A)
3. Look down at the model. Draw what you **observe**.

4. Position your eyes at table level. Look across the surface of the table at the model. Again, draw what you **observe**. (Picture B)

5. Look at the photographs at the bottom of page D55. One is of a spiral galaxy viewed from the edge. The other shows the galaxy viewed from the "top." **Compare** the pictures you drew in Steps 3 and 4 with the photographs of a spiral galaxy. Then look at page D54. **Observe** the photograph of a ribbon of stars. This is our view of the Milky Way Galaxy from Earth. Using your drawings and the photographs, **infer** where in the Milky Way Galaxy the sun is.

Picture A

Picture B

Draw Conclusions

1. Suppose the sun is located "above" the Milky Way Galaxy. What view of the galaxy might we see from Earth?
2. If the sun is in one of the arms, what view might we see of the galaxy then?
3. From your drawings and the photographs of a spiral galaxy, where in the Milky Way Galaxy do you **infer** the sun is?
4. **Scientists at Work** Scientists often **infer** when **using models** like the one you made. How did your model of the Milky Way Galaxy help you infer the sun's location in the galaxy?

Investigate Further **Observe** the Milky Way Galaxy in the night sky. You will need a clear, dark night, far away from city lights. Binoculars or a telescope will help you see some of the fainter stars.

Process Skill Tip

When you **infer** while **using models**, you base your inferences on everything you observe about the model. It's important to keep in mind that models are limited in how they show real-world situations.

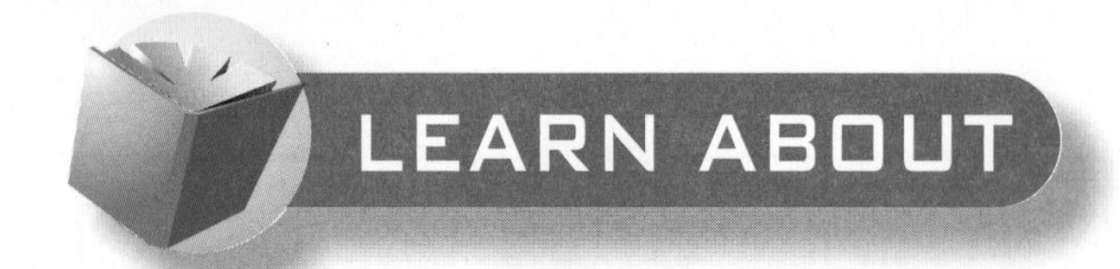

Galaxies

FIND OUT

- about the Milky Way Galaxy
- about types of galaxies
- about galactic clusters and nebulae

VOCABULARY

universe
galaxy
light-year

The Milky Way Galaxy

When you look at the sky on a clear, dark night, away from city lights, you can see millions of stars. But you are seeing only a small part of the universe. The **universe** is everything that exists—planets, stars, dust, gases, and energy. Although it seems crowded with stars, most of the universe is empty.

On some summer nights, you may see a bright ribbon of stars overhead. You are looking toward the center of Earth's galaxy. A **galaxy** is a group of stars, gas, and dust. Many galaxies rotate around a core. Most stars are part of galaxies, and the universe contains about a hundred billion galaxies.

Earth's galaxy is called the Milky Way Galaxy. It includes more than 100 billion stars, and it is one of the largest galaxies in the universe. It is so large that the light of a star on one side of the galaxy takes more than 100,000 light-years to reach the other side. As the Milky Way Galaxy rotates, the sun makes one complete turn around the center every 200–250 million Earth years.

✓ **What is a galaxy?**

This photograph of the Milky Way Galaxy was taken from Earth.

These drawings show the four types of galaxies. ►

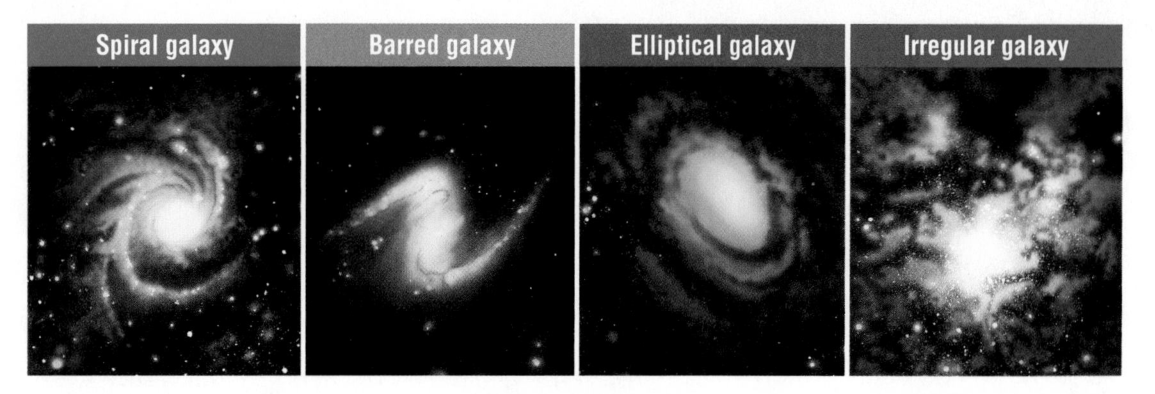

Types of Galaxies

Galaxies are classified by shape. There are four basic types: spiral, barred spiral, elliptical, and irregular.

The Milky Way Galaxy is a spiral galaxy. A spiral galaxy has a bright bulge of stars in the center and rotating arms. Earth's solar system is in one of the Milky Way Galaxy's spiral arms, about 30,000 light-years from the center of the galaxy. A **light-year** is the distance light travels in one Earth year, about 9.5 trillion km.

A spiral galaxy's arms contain young stars, protostars, dust, and gas. The thick bulge at the center contains older stars.

A barred spiral galaxy is similar to a spiral galaxy. The difference is that the spirals extend from a bar of stars that stretches from the center.

About half of all galaxies are elliptical. The shapes of elliptical galaxies range from almost a sphere, like a basketball, to a shape like a flattened football. Unlike spiral galaxies, elliptical galaxies don't seem to rotate. Irregular galaxies are groups of stars with no obvious shape.

✓ **What are the four basic shapes of galaxies?**

From the side a spiral galaxy looks like a thin rod with a center bulge. A spiral galaxy, seen from the "top," looks like a giant pinwheel spinning through space. The arms wind around the center as it turns, giving the galaxy its spiral appearance.

The Horsehead Nebula in Orion is a dark, dense swirl of dust. Behind the nebula, bright young stars flood the background with light.

Galactic Clusters and Nebulae

Have you ever seen a faint smudge or a misty patch of stars in the night sky? You may have been looking at a galactic cluster or a nebula.

A *galactic cluster* is a group of galaxies. The Milky Way Galaxy is one of about 30 galaxies in a cluster called the Local Group. The Milky Way Galaxy is one of the larger galaxies in this cluster. Most of the galaxies in the Local Group are small and elliptical or irregular.

Beyond the Local Group are other galactic clusters. Some of these are huge, with thousands of galaxies. One of the clusters closest to the Local Group is the Virgo Cluster, which is about 50 million light-years away.

You read about nebulae in Lesson 2. Astronomers hypothesize that stars form in nebulae. Unlike a galactic cluster, a nebula has no

◀ The Virgo Cluster contains more than 100,000 galaxies. At its center are three giant elliptical galaxies. One of these galaxies is about the same size as the entire Local Group.

light of its own. But if there is a hot star within a few light-years, the gases in the nebula shine. This is because hot stars emit a lot of ultraviolet waves. The energy of these waves is absorbed by the gases in the nebula. The gases then glow in different colors. The colors tell what gases are in the nebula. For example, hydrogen glows red, while oxygen and helium glow green. Different combinations of gases and dust create different effects, as in the Horsehead Nebula shown on page D56.

✓ **What is the difference between a galactic cluster and a nebula?**

Summary

A galaxy is a group of stars, gas, and dust. Many galaxies rotate around a central core. The sun is in the Milky Way Galaxy. The Milky Way Galaxy is part of a galactic cluster called the Local Group. Also visible in the night sky are nebulae, which are clouds of gas and dust in which stars form.

Review

1. What is the Local Group?
2. Describe the four basic shapes of galaxies.
3. What is a nebula?
4. **Critical Thinking** The Horsehead Nebula doesn't glow like some nebulae. Why not?
5. **Test Prep** The Virgo Cluster is a —
   - **A** giant elliptical galaxy
   - **B** nebula
   - **C** Local Group
   - **D** group of galaxies

LINKS

MATH LINK

Solve Problems It takes the sun about 250 million Earth years to travel around the center of the Milky Way Galaxy. If the sun is about 5 billion Earth years old, how many trips has it made?

WRITING LINK

Expressive Writing—Friendly Letter Suppose there are intelligent life forms in another galaxy. Write a letter to these life forms. Explain where you are in the Milky Way Galaxy, and tell the other life forms about life on Earth.

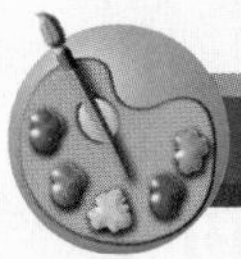

ART LINK

Galaxies on Stage Choose some of the information in this lesson. Rewrite it as a play. Present the play to other fifth-grade students in your school. Use dialogue, movement, and props in your play.

TECHNOLOGY LINK

Learn more about galaxies and how they move through space by viewing *Colliding Galaxies* on the **Harcourt Science Newsroom Video.**

Magnetars

Some people say "there's nothing new under the sun." Astronomers might laugh at this! They are always finding new things in the universe. One exciting discovery is a different kind of star. Stars of this kind spin rapidly and send out gigantic bursts of energy. They might also be the most powerful magnets in the universe. Astronomers call them magnetars.

This drawing of a magnetar shows its magnetic force field.

Surprise in the Sky

The first data about magnetars came from spy satellites. The U.S. government wanted to know when certain countries were testing nuclear weapons. So it launched a series of satellites to detect radiation from nuclear explosions. To government scientists' amazement, the satellites picked up radiation coming from outer space! Bursts of gamma rays seemed to be coming from everywhere in the universe. Government scientists communicated with civilian astronomers about this mysterious energy from space. Using satellites and radio telescopes, astronomers began measuring the energy bursts. Some of the larger bursts released as much energy in one second as the sun does in 1000 years!

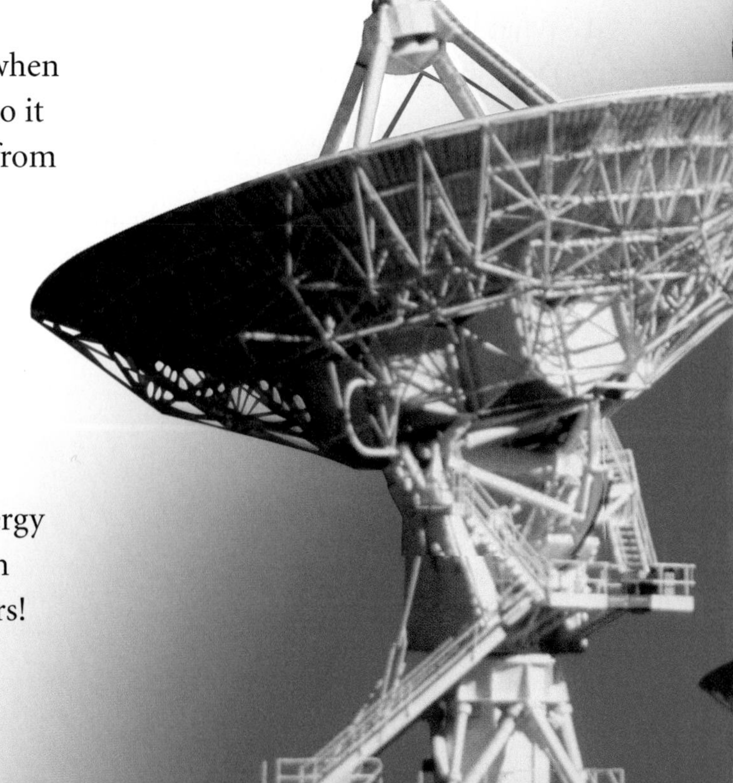

A few astronomers hypothesized that these stars have solid crusts and magnetic fields so strong that they cause the crusts to crack, releasing radiation. So magnetars may have starquakes, just as Earth has earthquakes.

A Giant Magnet

What's so different about this new kind of star? For one thing, if you were close enough, a magnetar's huge magnetic field would rearrange the atoms in your body. You wouldn't be you anymore! Think of this star as a giant magnet. Even if it were as far away as the sun, it could pull metal objects, such as paper clips, out of your pockets. The pull of gravity on its surface would flatten you like a pancake. And a magnetar spins very, very fast! While Earth takes 24 hours to complete one rotation, some magnetars take less than 6 seconds.

Radio telescopes are also used to identify unknown energy sources from space.

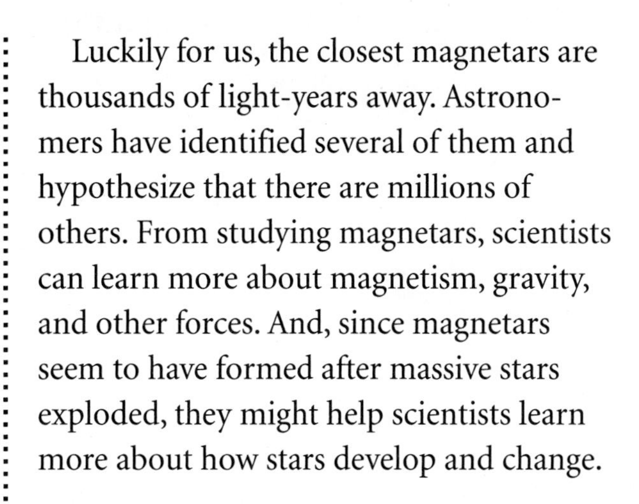

Luckily for us, the closest magnetars are thousands of light-years away. Astronomers have identified several of them and hypothesize that there are millions of others. From studying magnetars, scientists can learn more about magnetism, gravity, and other forces. And, since magnetars seem to have formed after massive stars exploded, they might help scientists learn more about how stars develop and change.

THINK ABOUT IT

1. How are magnetars different from other stars?
2. Why would it be dangerous for a spacecraft to fly near a magnetar?

CAREERS ASTROPHYSICIST

What They Do

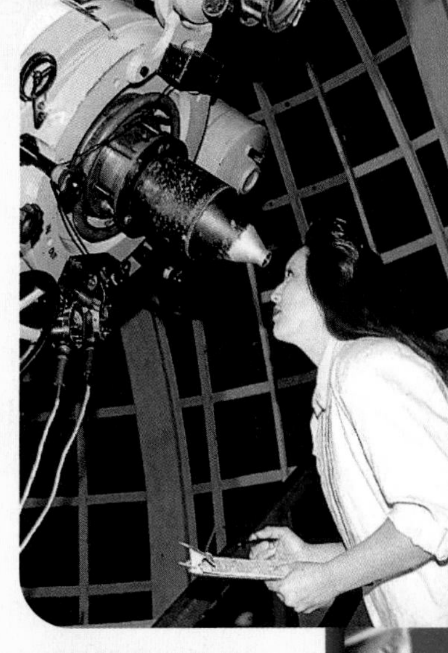

Astrophysicists work in space agencies, universities, or observatories. They study the stars and galaxies to find out more about radiation, magnetism, and other processes in the universe.

Education and Training

A person wishing to become an astrophysicist needs to study math, physics, and chemistry. He or she also needs training in using equipment such as radar and telescopes. Some astrophysicists learn how to use satellites to get data from distant stars.

WEB LINK
For Science and Technology updates, visit the Harcourt Internet site.
www.harcourtschool.com

Julio Navarro

ASTRONOMER

"If you look at a galaxy far away, you're seeing the galaxy as it was many years ago. You're looking at the universe as it was when it was young."

Julio Navarro is a theoretical astronomer who studies galaxies. Instead of looking through telescopes, Dr. Navarro works with numerical data from the Hubble Space Telescope and from ground-based telescopes. He uses this data in computer models that help him study the formation of galaxies and other bodies in the universe.

Dr. Navarro explains that the light from distant galaxies can take billions of years to reach our telescopes. So we're not seeing those galaxies as they appear now, but as they appeared billions of years ago.

When he was growing up in Argentina, no one would have guessed that Dr. Navarro would become an astronomer. He was more interested in music. However, an essay assignment in high school changed everything. "I was looking through a lot of books, and I became interested in astronomy," says Dr. Navarro. He kept reading, and his interest grew.

Dr. Navarro earned a bachelor's degree and a Ph.D. in astronomy before leaving Argentina to work and study in North America and Europe. Today Dr. Navarro is at the University of Victoria, in British Columbia, Canada, where he teaches astronomy and does research on the formation of galaxies.

THINK ABOUT IT

1. Why does Dr. Navarro study distant galaxies?
2. How does Dr. Navarro gather data for his studies?

ACTIVITIES FOR HOME OR SCHOOL

A MODEL SUN

What are some sun features?

Materials

- yellow construction paper
- ruler
- scissors
- markers
- black construction paper
- glue
- white legal-sized paper

Procedure

1. Cut a 20-cm circle from the yellow paper to represent the photosphere. Using markers, make sunspots on the photosphere.

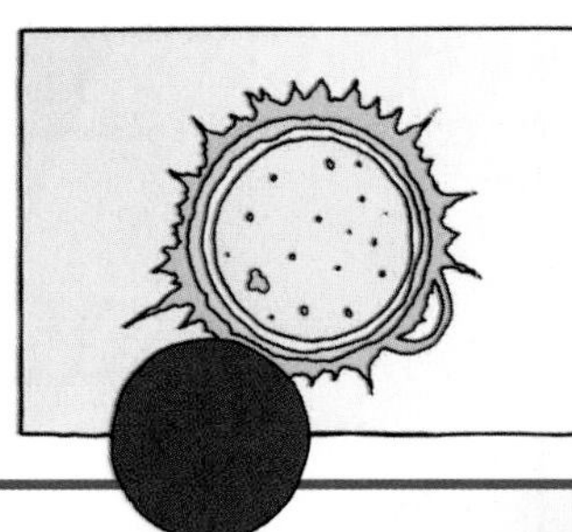

2. Cut a 20-cm circle from the black paper to represent the moon during a solar eclipse.
3. Glue the yellow circle to the white paper.
4. Using the markers, color jagged shapes around the sun to represent the sun's corona.
5. Use your black "moon" to eclipse the sun's photosphere to study its corona.

Draw Conclusions

Why did you cut out the moon for the total eclipse the same size as the sun? Scientists learn a great deal about the sun during total eclipses. Why is this an important time to study the sun? What benefits are there to blocking out the sun's photosphere?

ASTROLABE

How do people navigate by the stars?

Materials

- 15-cm cardboard square
- protractor
- pencil
- drinking straw
- tape
- 20-cm piece of string
- metal washer

Procedure

1. Using the protractor, and starting in one corner of the cardboard square, draw a line at an angle of 5°. Draw additional lines at 10°, 15°, and so on.
2. Tape the straw to the cardboard as shown.

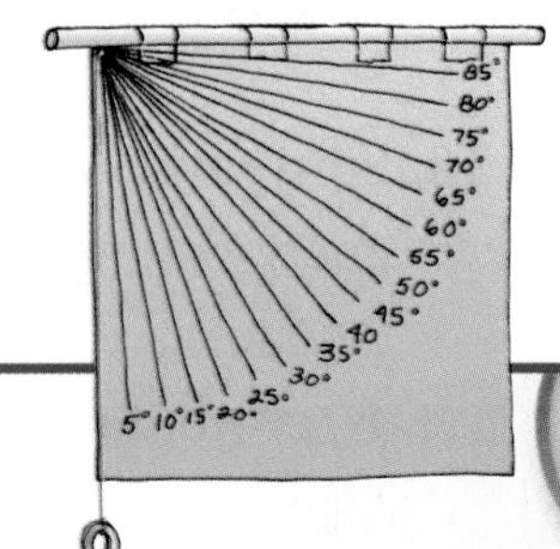

3. At the point where all the lines meet, make a hole in the cardboard. Push the string through the hole and tie a knot to keep it from pulling through the hole. Tie the washer to the other end of the string.
4. Look at the North Star through the straw. Measure the angle of the North Star by noting the angle of the string.

Draw Conclusions

The angle of the North Star tells you your latitude on the Earth. What is the angle of the North Star where you live? What is the latitude where you live?

CHAPTER 2 Review and Test Preparation

Vocabulary Review

Use the terms below to complete the sentences. The page numbers in ( ) tell you where to look in the chapter if you need help.

photosphere (D41)
corona (D41)
sunspot (D42)
solar flare (D42)
solar wind (D42)
magnitude (D46)
main sequence (D47)
universe (D54)
galaxy (D54)
light-year (D55)

1. The sun's atmosphere is the ____.
2. A dark area on the sun that is caused by twists and loops in the sun's magnetic field is a ____.
3. A ____ is a brief burst of energy that occurs above a sunspot.
4. The ____ is a fast-moving stream of particles ejected into space.
5. Two stars the same distance from Earth that give off the same amount of light will appear to have the same ____.
6. The sun's ____ is the layer we see.
7. The classification group to which most stars belong is the ____.
8. The ____ is everything that exists.
9. A group of stars, gas, and dust is a ____.
10. A ____ is the distance light travels in one Earth year.

Connect Concepts

Use the terms in the Word Bank below to complete the chart of Your Place in Space.

solar system **universe** **Earth**
Local Group **Milky Way Galaxy**

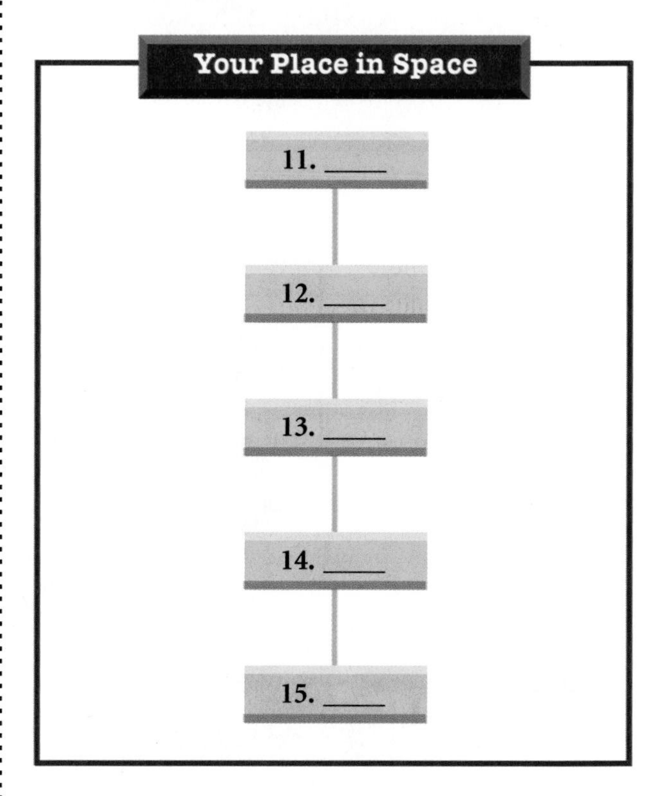

Check Understanding

Write the letter of the best choice.

16. At the center of the solar system is a —
    - **A** galaxy
    - **B** moon
    - **C** planet
    - **D** star
17. The surface of the sun is the part we can see. It is called the —
    - **F** core
    - **G** photosphere
    - **H** chromosphere
    - **J** corona
18. Sunspots can cause brief bursts of energy known as solar —
    - **A** flares
    - **B** prominences
    - **C** auroras
    - **D** fusion

19. The sun is a huge mass of —

F metals H gases
G light J liquids

20. The hottest stars in the main sequence are also the —

A brightest C largest
B least bright D smallest

21. The Milky Way Galaxy is located in the Local Group, a —

F universe H nebula
G galactic cluster J main sequence

22. What type of galaxy is shown in the illustration?

A elliptical C spiral
B irregular D spiral barred

Critical Thinking

23. Infrared telescopes have detected new stars forming in the Orion Nebula. Why can't we see these stars forming?

24. Galaxies contain millions of stars. Yet they look faint to people on Earth. Explain why galaxies look so faint to observers on Earth.

25. A light-year is the distance light travels in one Earth year. Suppose you look at a star that is 100,000 light-years away. Explain why you are looking into the past.

Process Skills Review

26. Sunspots appear and disappear in cycles that average about 11 years. As the cycle starts, the number of sunspots increases for 5 to 6 years. Then it decreases for 5 to 6 years. Using what you know about the relationship between solar flares and sunspots, **hypothesize** when solar flares are most frequent.

27. Scientists believe that time spent in the main sequence is one of the stages of a star's "life." Most known stars are main-sequence stars. From this information, what can you **infer** about the amount of time a typical star spends in this stage?

28. How could you **model** changes in the apparent magnitude of stars?

Performance Assessment

Star Light, Star Bright

Make up a star. Draw a picture of your star, showing its color and its type—main sequence, red giant, white dwarf, or nebula. Label the drawing with the type of star. Draw another picture, showing the location of your star in a spiral galaxy as viewed from "above." From your drawing, infer the direction in which the galaxy rotates. Draw an arrow to show the direction.

UNIT D EXPEDITIONS

There are many places where you can learn more about space. By visiting the places below, you can find out about the exploration of the solar system and beyond. You'll also have fun while you learn.

John C. Stennis Space Center

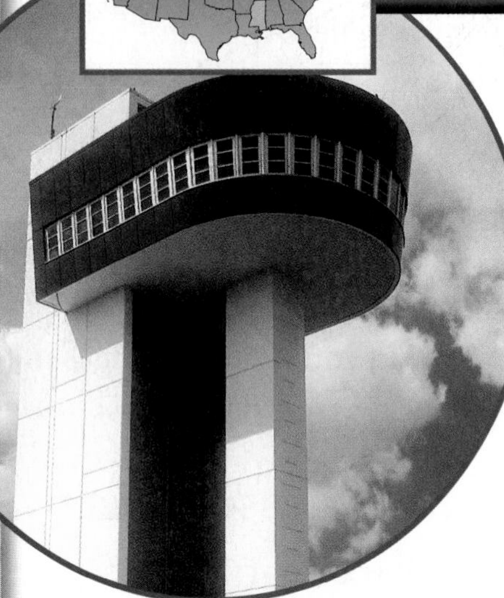

WHAT A collection of NASA spacecraft and exhibits about the exploration of space

WHERE I-10 Mississippi Welcome Station, Mississippi

WHAT CAN YOU DO THERE? See an Apollo-era Lunar Lander display, view other space exhibits, and try out the Space Shuttle and Mission to Mars simulators.

National Air and Space Museum

WHAT A museum with the world's largest collection of historic aircraft and spacecraft

WHERE Washington, D.C.

WHAT CAN YOU DO THERE? Tour the exhibits, see historic aircraft and spacecraft on display, and learn about the ongoing exploration of space.

GO ONLINE Plan Your Own Expeditions

If you can't visit the John C. Stennis Space Center or the National Air and Space Museum, visit a space exhibit or museum near you. Or log on to The Learning Site at **www.harcourtschool.com** to visit these science sites and other places where you can learn more about space.

References

Using Science Tools

Using a Hand Lens

A hand lens magnifies objects, or makes them look larger than they are.

1. Hold the hand lens about 12 centimeters (5 in.) from your eye.
2. Bring the object toward you until it comes into focus.

Using a Thermometer

A thermometer measures the temperature of air and most liquids.

1. Place the thermometer in the liquid. Don't touch the thermometer any more than you need to. Never stir the liquid with the thermometer. If you are measuring the temperature of the air, make sure that the thermometer is not in line with a direct light source.
2. Move so that your eyes are even with the liquid in the thermometer.
3. If you are measuring a material that is not being heated or cooled, wait about two minutes for the reading to become stable. Find the scale line that meets the top of the liquid in the thermometer, and read the temperature.
4. If the material you are measuring is being heated or cooled, you will not be able to wait before taking your measurements. Measure as quickly as you can.

Caring for and Using a Microscope

A microscope is another tool that magnifies objects. A microscope can increase the detail you see by increasing the number of times an object is magnified.

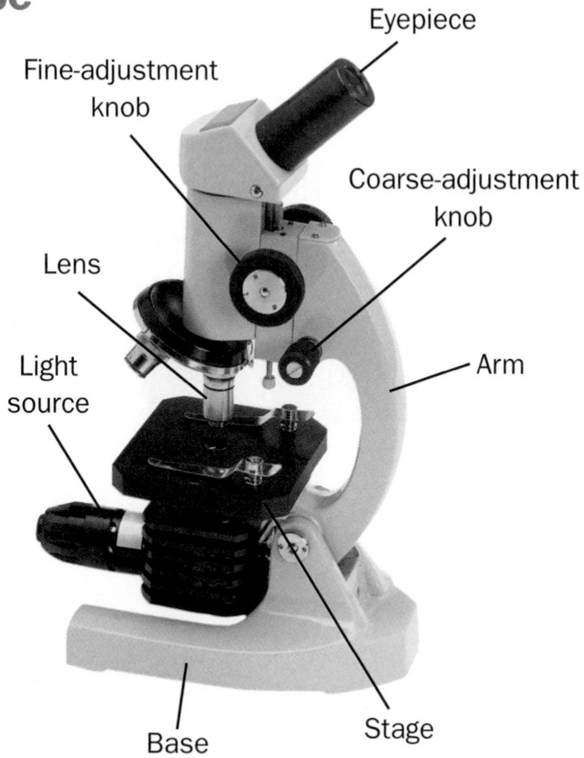

A Light Microscope

Caring for a Microscope

- Always use two hands when you carry a microscope.
- Never touch any of the lenses of a microscope with your fingers.

Using a Microscope

1. Raise the eyepiece as far as you can using the coarse-adjustment knob. Place your slide on the stage.
2. Always start by using the lowest power. The lowest-power lens is usually the shortest. Start with the lens in the lowest position it can go without touching the slide.
3. Look through the eyepiece, and begin adjusting it upward with the coarse-adjustment knob. When the slide is close to being in focus, use the fine-adjustment knob.
4. When you want to use a higher-power lens, first focus the slide under low power. Then, watching carefully to make sure that the lens will not hit the slide, turn the higher-power lens into place. Use only the fine-adjustment knob when looking through the higher-power lens.

You may use a Brock microscope. This is a sturdy microscope that has only one lens.

1. Place the object to be viewed on the stage.
2. Look through the eyepiece, and begin raising the tube until the object comes into focus.

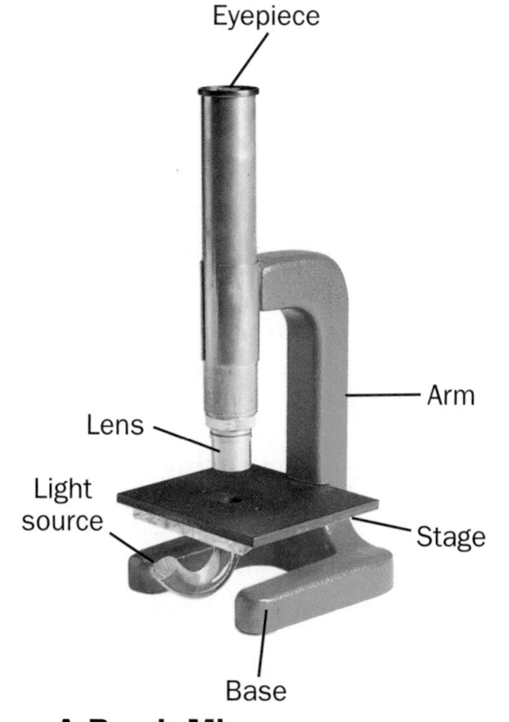

A Brock Microscope

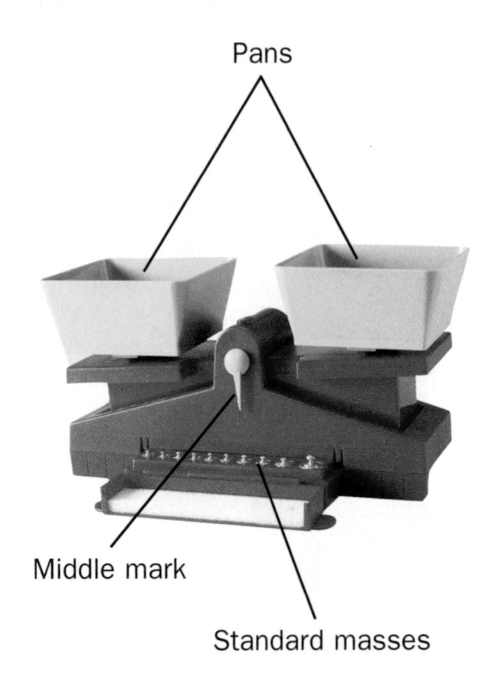

Using a Balance

Use a balance to measure an object's mass. Mass is the amount of matter an object has.

1. Look at the pointer on the base to make sure the empty pans are balanced.
2. Place the object you wish to measure in the left pan.
3. Add the standard masses to the other pan. As you add masses, you should see the pointer move. When the pointer is at the middle mark, the pans are balanced.
4. Add the numbers on the masses you used. The total is the mass in grams of the object you measured.

Using a Spring Scale

Use a spring scale to measure forces such as the pull of gravity on objects. You measure weight and other forces in units called newtons (N).

Measuring the Weight of an Object

1. Hook the spring scale to the object.
2. Lift the scale and object with a smooth motion. Do not jerk them upward.
3. Wait until any motion of the spring comes to a stop. Then read the number of newtons from the scale.

Measuring the Force to Move an Object

1. With the object resting on a table, hook the spring scale to it.
2. Pull the object smoothly across the table. Do not jerk the object.
3. As you pull, read the number of newtons you are using to pull the object.

Beaker Graduate

Measuring Liquids

Use a beaker, a measuring cup, or a graduate to measure liquids accurately.

1. Pour the liquid you want to measure into a measuring container. Put your measuring container on a flat surface, with the measuring scale facing you.
2. Look at the liquid through the container. Move so that your eyes are even with the surface of the liquid in the container.
3. To read the volume of the liquid, find the scale line that is even with the surface of the liquid.
4. If the surface of the liquid is not exactly even with a line, estimate the volume of the liquid. Decide which line the liquid is closer to, and use that number.

Using a Ruler or Meterstick

Use a ruler or meterstick to measure distances and to find lengths of objects.

1. Place the zero mark or end of the ruler or meterstick next to one end of the distance or object you want to measure.
2. On the ruler or meterstick, find the place next to the other end of the distance or object.
3. Look at the scale on the ruler or meterstick. This will show the distance you want or the length of the object.

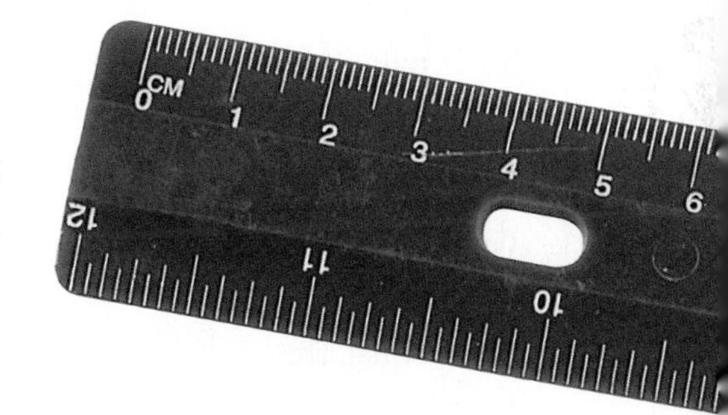

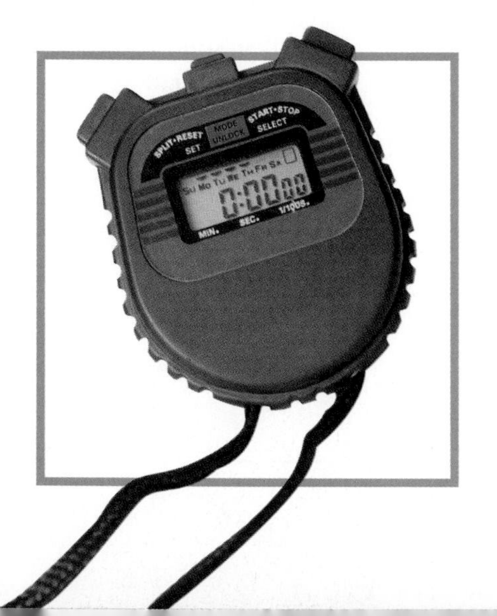

Using a Timing Device

Use a timing device such as a stopwatch to measure time.

1. Reset the stopwatch to zero.
2. When you are ready to begin timing, press start.
3. As soon as you are ready to stop timing, press stop.
4. The numbers on the dial or display show how many minutes, seconds, and parts of seconds have passed.

Visit the Multimedia Science Glossary to see illustrations of these words and to hear them pronounced.
www.harcourtschool.com/science

Glossary

As you read your science book, you will notice that new or unfamiliar words have been respelled to help you pronounce them quickly while you are reading. Those respellings are *phonetic respellings.* In this Glossary you will see a different kind of respelling. Here, diacritical marks are used, as they are used in dictionaries. *Diacritical respellings* provide a more precise pronunciation of the word.

When you see the ′ mark after a syllable, pronounce that syllable with more force than the other syllables. The page number at the end of the definition tells where to find the word in your book. The boldfaced letters in the examples in the Pronunciation Key that folows show how these letters are pronounced in the respellings after each glossary word.

PRONUNCIATION KEY

a	**a**dd, m**a**p	m	**m**ove, see**m**	u	**u**p, d**o**ne
ā	**a**ce, r**a**te	n	**n**ice, ti**n**	û(r)	b**ur**n, t**er**m
â(r)	c**a**re, **a**ir	ng	ri**ng**, so**ng**	yo͞o	f**u**se, f**ew**
ä	p**a**lm, f**a**ther	o	**o**dd, h**o**t	v	**v**ain, e**v**e
b	**b**at, ru**b**	ō	**o**pen, s**o**	w	**w**in, a**w**ay
ch	**ch**eck, cat**ch**	ô	**o**rder, j**aw**	y	**y**et, **y**earn
d	**d**og, ro**d**	oi	**oi**l, b**oy**	z	**z**est, mu**s**e
e	**e**nd, p**e**t	ou	p**ou**t, n**ow**	zh	vi**s**ion, plea**s**ure
ē	**e**qual, tr**ee**	o͝o	t**oo**k, f**u**ll	ə	the schwa, an
f	**f**it, hal**f**	o͞o	p**oo**l, f**oo**d		unstressed vowel
g	**g**o, lo**g**	p	**p**it, sto**p**		representing the sound
h	**h**ope, **h**ate	r	**r**un, poo**r**		spelled
i	**i**t, g**i**ve	s	**s**ee, pa**ss**		*a* in **a***bove*
ī	**i**ce, wr**i**te	sh	**s**ure, ru**sh**		*e* in *sick***e***n*
j	**j**oy, le**dge**	t	**t**alk, si**t**		*i* in *poss***i***ble*
k	**c**ool, ta**k**e	th	**th**in, bo**th**		*o* in *mel***o***n*
l	**l**ook, ru**l**e	t͟h	**th**is, ba**th**e		*u* in *circ***u***s*

Other symbols:

- • separates words into syllables
- ′ indicates heavier stress on a syllable
- ′ indicates light stress on a syllable

A

acceleration [ak•sel′ər•ā′shən] A change in motion caused by unbalanced forces or a change in velocity **(F13, F35)**

acid rain [as′id rān′] Precipitation resulting from pollution condensing into clouds and falling to Earth **(B99)**

action force [ak′shən fôrs′] The first force in the third law of motion **(F43)**

air mass [âr′ mas′] A large body of air that has nearly the same temperature and humidity throughout **(C75)**

air pressure [âr′ presh′ər] The weight of air **(C65)**

alveoli [al•vē′ə•lē] Tiny air sacs located at the ends of bronchi in the lungs **(A18)**

amphibians [am•fib′ē•ənz] Animals that have moist skin and no scales **(A44)**

angiosperm [an′jē•ō•spûrm′] A flowering plant **(A103)**

asexual reproduction [ā•sek′sho͞o•əl rē′prə•duk′shən] Reproduction by simple cell division **(A67)**

asteroids [as′tə•roidz] Chunks of rock that look like giant potatoes in space **(D16)**

atmosphere [at′məs•fir] The layer of air that surrounds Earth **(C64)**

atom [at′əm] The smallest unit of an element that has all the properties of that element **(E40)**

axis [ak′sis] An imaginary line that passes through Earth's center and its North and South Poles **(D7)**

B

balanced forces [bal′ənst fôrs′əz] The forces acting on an object that are equal in size and opposite in direction, canceling each other out **(F12)**

biomass [bī′ō•mas′] Organic matter, such as wood, that is living or was recently alive **(F110)**

biome [bī′ōm′] A large-scale ecosystem **(B64)**

birds [bûrdz] Vertebrates with feathers **(A45)**

bone marrow [bōn′ mar′ō] A connective tissue that produces red and white blood cells **(A24)**

capillaries [kap′ə•ler′ēz] The smallest blood vessels **(A17)**

carbon–oxygen cycle [kär′bən ok′sə•jən sī′kəl] The process by which carbon and oxygen cycle among plants, animals, and the environment **(B8)**

cell [sel] The basic unit of structure and function of all living things **(A6)**

cell membrane [sel′ mem′brān′] The thin covering that encloses a cell and holds its parts together **(A8)**

chemical bonds [kem′i•kəl bondz′] The forces that join atoms to each other **(F98)**

chlorophyll [klôr′ə•fil′] A pigment, or coloring matter, that helps plants use light energy to produce sugars **(A96)**

chromosome [krō′mə•sōm′] A threadlike strand of DNA inside the nucleus **(A65)**

classification [klas′ə•fə•kā′shən] The grouping of things by using a set of rules **(A38)**

climate [klī′mit] The average of all weather conditions through all seasons over a period of time **(C80)**

climate zone [klī′mit zōn′] A region throughout which yearly patterns of temperature, rainfall, and amount of sunlight are similar **(B64)**

climax community [klī′maks′ kə•myōō′nə•tē] The last stage of succession **(B93)**

combustibility [kəm•bus′tə•bil′ə•tē] The chemical property of being able to burn **(E24)**

comets [kom′its] Balls of ice and rock that circle the sun from two regions beyond the orbit of Pluto **(D16)**

community [kə•myōō′nə•tē] All the populations of organisms living together in an environment **(B28)**

competition [kom′pə•tish′ən] The contest among organisms for the limited resources of an ecosystem **(B42)**

compound [kom′pound] A substance made of the atoms of two or more different elements **(E48)**

condensation [kon′dən•sā′shən] The process by which a gas changes back into a liquid **(B14, C67, E17)**

conduction [kən•duk′shən] The direct transfer of heat between objects that touch **(F85)**

conductor [kən•duk′tər] A material that conducts electrons easily **(F70)**

conserving [kən•sûrv′ing] The saving or protecting of resources **(B104)**

consumer [kən•sōō′mər] An organism in a community that must eat to get the energy it needs **(B34)**

continental drift [kon′tə•nen′təl drift′] A theory of how Earth's continents move over its surface **(C22)**

convection [kən•vek′shən] The transfer of heat as a result of the mixing of a liquid or a gas **(F85)**

core [kôr] The center of the Earth **(C14)**

corona [kə•rō′nə] The sun's atmosphere **(D41)**

crust [krust] The thin, outer layer of Earth **(C14)**

current [kûr′ənt] A stream of water that flows like a river through the ocean **(C104)**

cytoplasm [sīt′ō•plaz′əm] A jellylike substance containing many chemicals that keep a cell functioning **(A9)**

decomposer [dē′kəm•pōz′ər] Consumer that breaks down the tissues of dead organisms **(B35)**

density [den′sə•tē] The concentration of matter in an object **(E9)**

deposition [dep′ə•zish′ən] The process of dropping, or depositing, sediment in a new location **(C7)**

desalination [dē•sal′ə•nā′shən] The process of removing salt from sea water **(C120)**

diffusion [di•fyōō′zhən] The process by which many materials move in and out of cells **(A10)**

direct development [də•rekt′ di•vel′əp•mənt] A kind of growth where organisms keep the same body features as they grow larger **(A72)**

dominant trait [dom′ə•nənt trāt′] A strong trait **(A79)**

earthquake [ûrth′kwāk′] A shaking of the ground caused by the sudden release of energy in Earth's crust **(C18)**

eclipse [i•klips′] The passing of one object through the shadow of another **(D8)**

ecosystem [ek′ō•sis′təm] A community and its physical environment together **(B28)**

electric charge [i•lek′trik chärj′] The charge obtained by an object when it gains or loses electrons **(F68)**

electric circuit [i•lek′trik sûr′kit] The path along which electrons can flow **(F71)**

electric current [i•lek′trik kûr′ənt] The flow of electrons from negatively charged objects to positively charged objects **(F69)**

electric force [i•lek′trik fôrs′] The attraction or repulsion of objects due to their charges **(F69)**

electromagnet [i•lek′trō•mag′nit] A temporary magnet made by passing electric current through a wire coiled around an iron bar **(F72)**

electron [ē•lek′tron′] A subatomic particle with a negative charge **(E39)**

element [el′ə•mənt] A substance made up of only one kind of atom **(E40)**

El Niño [el nēn′yō] A short-term climate change that occurs every two to ten years **(C83)**

endangered [en•dān′jərd] A term describing a population of organisms that is likely to become extinct if steps are not taken to save it **(B51)**

energy [en′ər•jē] The ability to cause changes in matter **(F62)**

energy pyramid [en′ər•jē pir′ə•mid] Shows the amount of energy available to pass from one level of a food chain to the next **(B38)**

equinox [ē′kwi•noks] Point in Earth's orbit at which the hours of daylight and darkness are equal **(D15)**

erosion [i•rō′zhən] The process of moving sediment from one place to another **(C7)**

estuary [es′chōō•er′ē] The place where a freshwater river empties into an ocean **(B80, C102)**

evaporation [ē•vap′ə•rā′shən] The process by which a liquid changes into a gas **(B14, C67, E16)**

exotic [ig•zot′•ik] An imported or nonnative organism **(B50)**

extinct [ik•stingkt′] No longer in existence; describes a species when the last individual of a population dies and that organism is gone forever **(B51)**

fault [fôlt] A break or place where pieces of Earth's crust move **(C18)**

fiber [fī′bər] Any material that can be separated into threads **(A112)**

fish [fish] Vertebrates that live their entire life in water **(A44)**

food chain [fōōd′ chān′] The ways in which the organisms in an ecosystem interact with one another according to what they eat **(B35)**

food web [fōōd′ web′] Shows the interactions among many different food chains in a single ecosystem **(B36)**

force [fôrs] A push or pull that causes an object to move, stop, or change direction **(F6)**

fossil [fos′əl] The remains or traces of past life found in sedimentary rock **(C23)**

friction [frik′shən] A force that opposes, or acts against, motion when two surfaces rub against each other **(F6)**

front [frunt] The boundary between air masses **(C75)**

fungi [fun′jī′] Living things that look like plants but cannot make their own food; example, mushrooms **(A39)**

fusion energy [fyōō′zhən en′ər•jē] The energy released when the nuclei of two atoms are forced together to form a larger nucleus **(F112)**

galaxy [gal′ək•sē] A group of stars, gas, and dust **(D54)**

gas [gas] The state of matter that does not have a definite shape or volume **(E14)**

gene [jēn] Structures on a chromosome that contain the DNA code for a trait an organism inherits **(A80)**

genus [jē′nas] The second-smallest name grouping used in classification **(A40)**

geothermal energy [jē′ō•thûr′məl en′ər•jē] Heat from inside the Earth **(F111)**

global warming [glō′bəl wôrm′ing] The hypothesized rise in Earth's average temperature from excess carbon dioxide **(C84)**

grain [grān] The seed of certain plants **(A110)**

gravitation [grav′i•tā′shən] The force that pulls all objects in the universe toward one another **(F8)**

greenhouse effect [grēn′hous′ i•fekt′] Process by which the Earth's atmosphere absorbs heat **(C84)**

gymnosperm [jim′nə•spûrm′] Plant with unprotected seeds; conifer or cone-bearing plant **(A102)**

habitat [hab′ə•tat′] A place in an ecosystem where a population lives **(B29)**

hardness [härd′nis] A mineral's ability to resist being scratched **(C37)**

headland [hed′land′] A hard, rocky point of land left when softer rock is washed away by the sea **(C111)**

heat [hēt] The transfer of thermal energy from one substance to another **(F84)**

humidity [hyōō•mid′ə•tē] A measure of the amount of water in the air **(C65)**

hydroelectric energy [hī′drō•ē•lek′trik en′ər•jē] Electricity generated from the force of moving water **(F104)**

igneous rock [ig′nē•əs rok′] A type of rock that forms when melted rock hardens **(C42)**

individual [in′də•vij′ōō•əl] A single organism in an environment **(B28)**

inertia [in•ûr′shə] The property of matter that keeps it moving in a straight line or keeps it at rest **(F41)**

inherited trait [in•her′it•əd trāt′] A characteristic that is passed from parent to offspring **(A78)**

instinct [in′stingkt] A behavior that an organism inherits **(B46)**

insulator [in′sə•lāt′ər] A material that does not carry electrons **(F71)**

intertidal zone [in′tər•tīd′əl zōn′] An area where the tide and churning waves provide a constant supply of oxygen and nutrients to living organisms **(B77)**

invertebrates [in•vûr′tə•brits] Animals without a backbone **(A45)**

jetty [jet′ē] A wall-like structure made of rocks that sticks out into the ocean **(C112)**

joint [joint] A place where bones meet and are attached to each other and to muscles **(A24)**

kinetic energy [ki•net′ik en′ər•jē] The energy of motion, or energy in use **(F62)**

kingdom [king′dəm] The largest group into which living things can be classified **(A39)**

landform [land′fôrm′] A physical feature on Earth's surface **(C6)**

law of universal gravitation [lô′ uv yōōn′ə•vûr′səl grav′i•tā′shən] Law that states that all objects in the universe are attracted to all other objects **(F49)**

learned behavior [lûrnd′ bē•hāv′yər] A behavior an animal learns from its parents **(B46)**

lens [lenz] A piece of clear material that bends, or refracts, light rays passing through it **(F77)**

ligament [lig′ə•mənt] One of the bands of connective tissue that hold a skeleton together **(A25)**

light-year [līt′yir′] The distance light travels in one Earth year; about 9.5 trillion km **(D55)**

liquid [lik′wid] The state of matter that has a definite volume but no definite shape **(E14)**

local winds [lō′kəl windz′] The winds dependent upon local changes in temperature **(C73)**

luster [lus′tər] The way the surface of a mineral reflects light **(C37)**

machine [mə•shēn′] Something that makes work easier by changing the size or the direction of a force **(F20)**

magma [mag′mə] Hot, soft rock from Earth's lower mantle **(C16)**

magnetism [mag′nə•tiz′əm] The force of repulsion (pushing) or attraction (pulling) between poles of magnets **(F7)**

magnitude [mag′nə•tōōd] Brightness of stars **(D46)**

main sequence [mān′ sē′kwəns] A band of stars that includes most stars of average color, size, magnitude, and temperature **(D47)**

mammals [mam′əlz] Animals that have hair and produce milk for their young **(A44)**

mantle [man′təl] The layer of rock beneath Earth's crust **(C14)**

mass [mas] The amount of matter in an object **(E7)**

mass movement [mas′ mōōv′mənt] The downhill movement of rock and soil because of gravity **(C9)**

matter [mat′ər] Anything that has mass and takes up space **(E6)**

meiosis [mī•ō′sis] The process that reduces the number of chromosomes in reproductive cells **(A68)**

metamorphic rock [met′ə•môr′fik rok′] A type of rock changed by heat or pressure but not completely melted **(C46)**

metamorphosis [met′ə•môr′fə•sis] A change in the shape or characteristics of an organism's body as it grows **(A73)**

microclimate [mī′krō•klī′mit] The climate of a very small area **(C80)**

mineral [min′ər•əl] A natural, solid material with particles arranged in a repeating pattern **(C36)**

mitosis [mī•tō′sis] The process of cell division **(A65)**

molecule [mol′ə•kyōōl′] A grouping of two or more atoms joined together **(E40)**

moneran [mō•ner′ən] The kingdom of classification for organisms that have only one cell and no nucleus **(A39)**

momentum [mō•men′təm] A measure of how hard it is to slow down or stop an object **(F36)**

near-shore zone [nir′shôr′ zōn′] The area beyond the breaking waves that extends to waters that are about 180 m deep **(B77)**

nephrons [nef′ronz′] Tubes inside the kidneys where urea and water diffuse from the blood **(A20)**

net force [net′ fôrs′] The result of two or more forces acting together on an object **(F14)**

neuron [nŏŏr′on′] A specialized cell that can receive information and transmit it to other cells **(A26)**

neutron [nōō′tron′] A subatomic particle with no charge **(E39)**

niche [nich] The role each population has in its habitat **(B29)**

nitrogen cycle [nī′trə•jən sī′kəl] The cycle in which nitrogen gas is changed into forms of nitrogen that plants can use **(B7)**

nonvascular plants [non•vas′kyə•lər plants] Plants that do not have tubes **(A52)**

nuclear energy [nōō′klē•ər en′ər•jē] The energy released when the nucleus of an atom is split apart **(F110)**

nucleus [nōō′klē•əs] **1** *(cell)* The organelle that controls all of a cell's activities **2** *(atom)* The center of an atom **(A8, E39)**

open-ocean zone [ō′pən•ō′shən zōn′] The area that includes most deep ocean waters; most organisms live near the surface **(B77)**

orbit [ôr′bit] The path one body in space takes as it revolves around another body; such as that of Earth as it revolves around the sun **(D7, F48)**

organ [ôr′gən] Tissues that work together to perform a specific function **(A12)**

osmosis [os•mō′sis] The diffusion of water and dissolved materials through cell membranes **(A10)**

Pangea [pan•jē′ə] A supercontinent containing all of Earth's land that existed about 225 million years ago **(C22)**

periodic table [pir′ē•od′ik tā′bəl] The table of elements in order of increasing atomic number; grouped by similar properties **(E47)**

phloem [flō′em] The tubes that transport food in the vascular plants **(A95)**

photosphere [fōt′ə•sfir′] The visible surface of the sun **(D41)**

photosynthesis [fōt′ō•sin′thə•sis] The process by which plants make food **(A96)**

physical properties [fiz′i•kəl prop′ər•tēz] The characteristics of a substance that can be observed or measured without changing the substance **(E6)**

pioneer plants [pī′ə•nir′ plantz′] The first plants to invade a bare area **(B92)**

pitch [pich] An element of sound determined by the speed at which sound waves move **(F79)**

planets [plan′its] Large, round bodies that revolve around a star **(D16)**

plate [plāt] A rigid block of crust and upper mantle rock **(C15)**

pollen [pol′ən] Flower structures that contain the male reproductive cells **(A102)**

pollution [pə•lo͞o′shən] Waste products that damage an ecosystem **(B99)**

population [pop•yə•lā′shən] All the individuals of the same kind living in the same environment **(B28)**

position [pə•zish′ən] An object's place, or location **(F34)**

potential energy [pō•ten′shəl en′ər•jē] The energy an object has because of its place or its condition **(F62)**

power [pou′ər] The amount of work done for each unit of time **(F19)**

precipitation [pri•sip′ə•tā′shən] Any form of water that falls from clouds, such as rain or snow **(B15, C65)**

prevailing winds [prē•vāl′ing windz′] The global winds that blow constantly from the same direction **(C73)**

producer [prə•do͞os′ər] An organism that makes its own food **(B34)**

protist [prō′tist] The kingdom of classification for organisms that have only one cell and also have a nucleus, or cell control center **(A39)**

proton [prō′ton′] A subatomic particle with a positive charge **(E39)**

radiation [rā′dē•ā′shən] The transfer of thermal energy by electromagnetic waves **(F85)**

reaction force [rē•ak′shən fôrs′] The force that pushes or pulls back in the third law of motion **(F43)**

reactivity [rē′ak•tiv′ə•tē] The ability of a substance to go through a chemical change **(E23)**

receptors [ri•sep′tərz] Nerve cells that detect conditions in the body's environment **(A26)**

recessive trait [ri•ses′iv trāt′] A weak trait **(A79)**

reclamation [rek′lə•mā′shən] The process of restoring a damaged ecosystem **(B110)**

recycle [rē•sī′kəl] To recover a resource from an item and use the recovered resource to make a new item **(B105)**

reduce [ri•do͞os′] To cut down on the use of resources **(B104)**

reflection [ri•flek′shən] The light energy that bounces off objects **(F76)**

refraction [ri•frak′shən] The bending of light rays when they pass through a substance **(F76)**

reptiles [rep′tīlz] Animals that have dry, scaly skin **(A44)**

resistor [ri•zis′tər] A material that resists the flow of electrons in some way **(F71)**

respiration [res′pə•rā′shən] The process that releases energy from food **(B8)**

reuse [rē′yo͞oz′] To use items again, sometimes for a different purpose **(B105)**

revolve [ri•volv′] To travel in a closed path around an object such as Earth does as it moves around the sun **(D6)**

rock [rok] A material made up of one or more minerals **(C42)**

rock cycle [rok′ sī′kəl] The slow, never-ending process of rock changes **(C52)**

rotate [rō′tāt] The spinning of Earth on its axis **(D7)**

S

salinity [sə•lin′ə•tē] Saltiness of the ocean **(C97)**

satellite [sat′ə•līt′] A natural body, like the moon, or an artificial object that orbits another object **(D23)**

scuba [sko͞o′bə] Underwater breathing equipment; the letters stand for **s**elf-**c**ontained **u**nderwater **b**reathing **a**pparatus **(C117)**

sedimentary rock [sed′ə•men′tər•ē rok′] A type of rock formed by layers of sediments that were squeezed and stuck together over a long time **(C44)**

sexual reproduction [sek′sho͞o•əl rē′prə•duk′shən] The form of reproduction in which cells from two parents unite to form a zygote **(A68)**

shore [shôr] The area where the ocean and land meet and interact **(C110)**

solar energy [sō′lər en′ər•jē] The energy of sunlight **(F111)**

solar flare [sō′lər flâr′] A brief burst of energy from the sun's photosphere **(D42)**

solar wind [sō′lər wind′] A fast-moving stream of particles thrown into space by solar flares **(D42)**

solid [sol′id] The state of matter that has a definite shape and a definite volume **(E14)**

solstice [sol′stis] Point in Earth's orbit at which the hours of daylight are at their greatest or fewest **(D15)**

solubility [sol′yə•bil′ə•tē] The ability of one substance to be dissolved in another substance **(E10)**

sonar [sō′när′] A device that uses sound waves to determine water depth **(C117)**

space probe [spās′ prōb′] A robot vehicle used to explore deep space **(D24)**

species [spē′shēz] The smallest name grouping used in classification **(A40)**

speed [spēd] A measure of the distance an object moves in a given amount of time **(F35)**

spore [spôr] A single reproductive cell that grows into a new plant **(A101)**

streak [strēk] The color of the powder left behind when you rub a material against a white tile called a streak plate **(C37)**

submersible [sub•mûr′sə•bəl] An underwater vehicle **(C117)**

succession [sək•sesh′ən] A gradual change in an ecosystem, sometimes occurring over hundreds of years **(B92)**

sunspot [sun′spot′] A dark spot on the photosphere of the sun **(D42)**

symbiosis [sim′bē•ō′sis] A long-term relationship between different kinds of organisms **(B45)**

system [sis′təm] Organs that work together to perform a function **(A12)**

telescope [tel′ə•skōp′] An instrument that magnifies distant objects, or makes them appear larger **(D23)**

temperature [tem′pər•ə•chər] The average kinetic energy of all the molecules in an object **(F84)**

tendons [ten′dənz] Tough bands of connective tissue that attach muscles to bones **(A25)**

threatened [thret′ənd] Describes a population of organisms that are likely to become endangered if they are not protected **(B51)**

tidal energy [tīd′əl en′ər•jē] A form of hydroelectric energy that produces electricity from the rising and falling of tides **(F106)**

tide [tīd] The repeated rise and fall in the level of the ocean **(C106)**

tide pool [tīd′ pōōl′] A pool of sea water found along a rocky shoreline **(C111)**

tissue [tish′ōō] Cells that work together to perform a specific function **(A12)**

transpiration [tran′spə•rā′shən] The process in which plants give off water through their stomata **(B15)**

unbalanced forces [un•bal′ənst fôrs′əz] Forces that are not equal **(F13)**

universe [yōōn′ə•vûrs′] Everything that exists—planets, stars, dust, gases, and energy **(D54)**

vascular plants [vas′kyə•lər plants] Plants that have tubes **(A50)**

velocity [və•los′ə•tē] An object's speed in a particular direction **(F35)**

vertebrates [vûr′tə•brits] Animals with a backbone **(A44)**

villi [vil′ī] Projections sticking into the small intestine **(A19)**

volcano [vol•kā′nō] A mountain formed by lava and ash **(C16)**

volume [vol′yōōm] **1** *(measurement)* The amount of space that an object takes up **2** *(sound)* The loudness of a sound **(E8, F79)**

water cycle [wôt′ər sī′kəl] The cycle in which Earth's water moves through the environment **(B14)**

water pressure [wôt′ər presh′ər] The weight of water pressing on an object **(C97)**

wave [wāv] An up-and-down movement of surface water **(C102)**

weathering [weth′ər•ing] The process of breaking rock into soil, sand, and other tiny pieces **(C7)**

weight [wāt] A measure of the pull of gravity on an object **(E7)**

wetlands [wet′landz′] The water ecosystems that include saltwater marshes, mangrove swamps, and mud flats **(B111)**

work [wûrk] The use of a force to move an object through a distance **(F18)**

xylem [zī′ləm] The tubes that transport water and minerals in vascular plants **(A95)**

T

Page Placement Key:
(l)-left, (r)-right, (t)-top, (c)-center, (b)-bottom, (bg)-background, (fg)-foreground, (i)-inset

Cover and Title Pages
Wolfgang Kaehler/Corbis; (bg) Eduardo Garcia/FPG International

Table of Contents
iv (t) Anup & Manoj Shah/Animals Animals; iv (bg) Grant V. Faint/The Image Bank; v (t) Zig Leszczynski/Animals Animals; v (bg) Karl Hentz/The Image Bank; vi (t) Eric & David Hosking/Photo Researchers; vi (bg) Bios (Klein-Hubert)/Peter Arnold, Inc; vii Telegraph Colour Library/FPG International; viii (t) Alvis Upitis/The Image Bank; viii (bg) Tim Crosby/Liaison International; ix (t) David Zaitz/Photonica; ix (bg) Stone;

Unit A
Unit A Opener (fg) Anup & Manoj Shah/Animals Animals; (bg)Grant V. Faint/The Image Bank; A2-A3 Image Shop/Phototake; A3 (l) Lawrence Migdale/Photo Researchers; A3 (c) Quest/Science Photo Library/Photo Researchers; A4 Charles D. Winters/Timeframe Photography, Inc./Photo Researchers; A6 (l) The Granger Collection, New York; A6 (c), (r) Courtesy of Hunt Institute for Botanical Documentation, Carnegie Mellon University, Pittsburgh, PA; A7 (tl) Ed Reschke/Peter Arnold, Inc.; A7 (tr) Michel Viard/Peter Arnold, Inc.; A7 (bl) Courtesy of Dr. Sam Harbo D.V.M., and Dr. Jurgen Schumacher D.V.M. , Veterinary Hospital, University of Tennessee; A7 (br) A.B. Sheldon/Dembinsky Photo Associates; A8 Dwight R. Kuhn; A9 Courtesy of Dr. Sam Harbo D.V.M., and Dr. Jurgen Schumacher D.V.M., Veterinary Hospital, University of Tennessee; A11 Skip Moody/Dembinsky Photo Associates; A14 Michael Newman/PhotoEdit; A16 (l) Dr. Tony Brain/Science Photo Library/Photo Researchers; A16 (r) Prof. P. Motta/Dept. of Anatomy/University "La Sapienza", Rome/Science Photo Library/Photo Researchers; A22 Gary Holscher/Stone; A28 D. Cavagnaro/DRK; A28 (i) Dr. Dennis Kunkel/Phototake; A29 Mark Richards/PhotoEdit; A30 NASA; A33 Charles D. Winters/Timeframe Photography, Inc./Photo Researchers; A34-A35 Gregory Ochocki/Photo Researchers; A35 (t) Dave Watts/Tom Stack & Associates; A35 (b) Frances Fawcett/Cornell University/American Indian Program; A36 Christian Grzimek/Okapia/Photo Reseachers; A38-A39 Bill Lea/Dembinsky Photo Associates; A38 (l) MESZA/Bruce Coleman, Inc.; A38 (c) Andrew Syred/SPL/Photo Researchers; A38 (r) Robert Brons/BPS/Stone; A39 (t) Bill Lea/Dembinsky Photo Associates; A39 (tc) Dr. E. R. Degginger/Color-Pic; A39 (c) S. Nielsen/Bruce Coleman, Inc.; A39 (bc) Robert Brons/BPS/Stone; A39 (b) Andrew Syred/SPL/Photo Researchers; A41 Daniel Cox/Stone; A42 Arthur C. Smith, III/Grant Heilman Photography; A44 (t) Ana Laura Gonzalez/Animals Animals; A44 (b) Tom Brakefield/The Stock Market; A44-A45 Runk/Schoenberger/Grant Heilman Photography; A45 (tl) Amos Nachoum/The Stock Market; A45 (tc) Hans Pfletschinger/Peter Arnold, Inc.; A45 (tr) Mark Moffett/Minden Pictures; A45 (br) Larry Lipsky/DRK; A46 (t) James Balog/Stone; A46 (b) Stephen Dalton/Photo Researchers; A48 Darrell Gulin/Stone; A50 Dr. E. R. Degginger, FPSA/Color-Pic; A51 Phil A. Dotson/Photo Researchers; A52 (t) Heather Angel/Biofotos; A52 (c) Runk Schoenberger/Grant Heilman Photography; A52-A53 Runk Schoenberger/Grant Heilman Photography; A54 Leonard Lee Rue III/Photo Researchers; A54-A55 S. J. Krasemann/Peter Arnold, Inc.; A55 (tl) Art Resource, NY; A55 (tr) Dr. E. R. Degginger/Photo Researchers; A55 (bl) Superstock; A55 (br) The Granger Collection, New York; A56 (t) Courtesy of Hunt Institute for Botanical Documentation, Carnegie Mellon University, Pittsburg, PA; A56 (b) Grant Heilman Photography; A60-A61 Rob & Ann Simpson/Visuals Unlimited; A61 (l) Dwight R. Kuhn; A61 (r) Dr. D. Spector/Peter Arnold, Inc.; A62 Ron Kimball; A64 (l) Jerome Wexler/Photo Researchers; A64 (cl), (c) Carolina Biological Supply Company/Phototake; A64 (cr) Jerome Wexler/Photo Researchers; A64 (r) Kenneth H. Thomas/Photo Researchers; A65 Conly L. Rieder/BPS/Stone; A66 (tl), (tc), (tr) Carolina Biological Supply Company/Phototake; A66 (c) Noble Proctor/Photo Researchers; A66 (b) Zig Leszczynski/Animals Animals; A67 (tl), (tc), (tr) Carolina Biological Supply Company/Phototake; A67 (b) Bob Gossington/Bruce Coleman, Inc.; A69 Carolina Biological Supply Company/Phototake; A70 J.H. Robinson/Photo Researchers; A72 (t) Peter A. Simon/Phototake; A72 (b) Dr. E.R. Degginger/Photo Researchers; A73 (l) Thomas Gulz/Visuals Unlimited; A73 (c) Dwight R. Kuhn; A73 (r) William J. Weber/Visuals Unlimited; A74 Harry Rogers/Photo Researchers; A75 Michael Fogden/Bruce Coleman, Inc.; A76 Paul Barton/The Stock Market; A78 Phil Savoie/The Picture Cube; A79 The Granger Collection, New York; A82 Tim Davis/Tony Stone Images; A83 (li) College of Veterinary Medicine/University of Florida; A83 (r) Zigy Kaluzny/Tony Stone Images; A84 Henry Friedman/HRW; A84 (i) Oliver Meckes/Photo Researchers; A88-A89 Tom Bean/Stone; A89 (t) Inga Spence/Visuals Unlimited; A89 (b) Ned Therrien/Visuals Unlimited; A90 James Randklev/Stone; A92 (l) Richard Choy/Peter Arnold, Inc.; A92 (r) Reinhard Siegel/Stone; A93 Norman Myers/Bruce Coleman, Inc.; A93 (li) Dr. E. R. Degginger/Color-Pic; A93 (ri) John Kaprielian/Photo Researchers; A95 Jane Grushow/Grant Heilman Photography; A96-A97 (t), A96 (ti) Runk/Schoenberger/Grant Heilman Photography; A96-A97 (b) Alan Levenson/Stone; A98 Darrell Gulin/Dembinsky Photo Associates; A100 Kim Taylor/Bruce Coleman, Inc.; A101, A102 (t) Runk/Schoenberger/Grant Heilman Photography; A102 (b) S.J. Krasemann/Peter Arnold, Inc.; A103 (t) Dr. E. R. Degginger/Color-Pic; A103 (b) Robert Maier/Earth Scenes; A104 (t) David Cavagnaro/Peter Arnold, Inc.; A104 (tc) E. R. Degginger/Bruce Coleman, Inc.; A104 (bc) Gregory K. Scott/Photo Researchers; A104 (b) Kevin Schafer Photography; A104 (bg) Jeff Lepore/Photo Researchers; A105 Runk/Schoenberger/Grant Heilman Photography; A106 (animal life cycle) (t) Gregory K. Scott/Photo Researchers; A106 (r) Harry Rogers/National Audubon Society; A106 (b) David M. Dennis/Tom Stack & Associates; A106 (l) Jen & Des Bartlett/Bruce Coleman, Inc.; A106 (plant life cycle) (t) Dr. E. R. Degginger/Color-Pic; A106 (r) Barry L. Runk/Grant Heilman Photography; A106 (b) Jane Grushow/Grant Heilman Photography; A106 (l) Dwight R. Kuhn; A112 (l) Alan & Linda Detrick/Photo Researchers; A112 (cr) Angelina Lax/Photo Researchers; A113 Grant Heilman Photography; A113 (i) Will & Deni MvIntyre/Photo Researchers; A114-115 Dana Downie/AGStock USA; A115 (b) Mark Richards/PhotoEdit; A116 Dennis Carlyle Darling/ HRW; A118 Dr. E. R. Degginger/Color-Pic; A119 James Randklev/Stone; A120 (t) Jeff Greenberg/Unicorn Stock Photos; A120 (b) Jack Olson Photography;

Unit B
Unit B Opener (fg) Zig Leszczynski/Animals Animals; (bg) Karl Hentz/The Image Bank; B2 Clyde H. Smith/Peter Arnold, Inc.; B2-B3 Superstock; B3 Earl Roberge/Photo Researchers; B4 Wolfgang Kaehler Photography; B6 Randy Ury/The Stock Market; B7 Thomas Hovland/Grant Heilman Photography; B10 Wolfgang Kaehler Photography; B12 Michael Giannechini/Photo Researchers; B14-B15 Greg Vaughn/Stone; B16 (t) C. Vincent/Natural Selection Stock Photography; B16 (b) Bob Daemmrich Photography, Inc.; B16 (bi) Superstock; B18 John Shaw/Bruce Coleman, Inc.; B18-B19 Lee Rentz/Bruce Coleman, Inc.; B19 Ken Graham/Bruce Coleman, Inc.; B20 Sipa Press; B22 Greg Vaughn/Stone; B23 C. Vincent/Natural Selection Stock Photography; B24-B25 P & R Hagan/Bruce Coleman, Inc.; B25 (t) Tomas del Amo/Pacific Stock; B25 (b) Mitsuaki Iwago/Minden Pictures; B26 Tim Davis/Photo Researchers; B28 (li) Michael Giannechini/Photo Researchers; B28-B29 (bg) J.A. Kravlis/Masterfile; B28-B29 (ci) Ted Kerasote/Photo Researchers; B29 (ti) Superstock; B29 (bi) Mitsuaki Iwago/Minden Pictures; B30 (tl) David Muench Photography, Inc.; B30 (tr), (bl) Barry L. Runk/Grant Heilman Photography; B30 (br) David Muench Photography, Inc.; B32 Superstock; B34 (tli) V.P. Weinland/Photo Researchers; B34 (tri) Parviz M. Pour/Photo Researchers; B34-B35 (bi) Dembinsky Photo Associates; B34-B35 (bg) Larry Ditto/Bruce Coleman, Inc.; B35 (li) Tom McHugh/Photo Researchers; B35 (ri) Tom & Pat Leeson/Photo Researchers; B36-B37 Woods, Michael J./NGS Image Collection; B39 Bruce Coleman, Inc.; B40 (both) Joe McDonald/McDonald Wildlife Photography; B42 (bg) Stuart Westmorland/Stone; B42 (li) Roger Bickel/New England Stock Photo; B42 (ri) Bruce Coleman, Inc.; B43 (l) Kevin Schafer/Peter Arnold, Inc.; B43 (r) Mitsuaki Iwago/Minden Pictures; B44 (t) John Shaw/Bruce Coleman, Inc.; B44 (c) Hal H. Harrison/Photo Researchers; B44 (b) Wayne Lankinen/Bruce Coleman, Inc.; B45 (t) M. & C. Photography/Peter Arnold, Inc.; B45 (b) William Townsend/Photo Researchers; B46 (t) Vince Streano/The Stock Market; B46-B47 Ralph Ginzburg/Peter Arnold, Inc.; B48 Bryan & Cherry Alexander/Masterfile; B50 (t) Tim Davis/Photo Researchers; B50 (bl) Johnny Johnson/Tony Stone Images; B50 (br) Malcolm Boulton/Photo Researchers; B51 Tom McHugh/Photo Researchers; B52-B53 Ted Schiffman/Peter Arnold, Inc; B52 Roy Toft/Tom Stack & Associates; B54 Gunter Ziesler/Peter Arnold, Inc.; B55 (t) Doug Cheeseman/Peter Arnold, Inc.; B55 (b) Bonnie Kamin/PhotoEdit; B56 (i) Louisiana State University Chemistry Library Website; B56 Meckes/Ottawa/Photo Researchers; B60-B61 Craig Tuttle/The Stock Market; B61 (t) Jake Rajs/Stone; B61 (b) Earth Satellite Corporation/Science Photo Library/Photo Researchers; B62 Chromosohm/Sohm/Stone; B64 (t) David Muench Photography, Inc.; B64 (b) Gary Braasch/Stone; B65 (tl) Superstock; B65 (tr) Steve Kaufman/Peter Arnold, Inc.; B65 (bl) Joseph Van Os/The Image Bank; B65 (br) Colin Prior/Stone; B66 Wolfgang Kaehler Photography; B66 (i) Mark Moffett/Minden Pictures; B67 Superstock; B67 (i) Roger Bickel/New England Stock Photo; B68 David Muench Photography, Inc.; B68 (i) William Manning/The Stock Market; B69 Darrell Gulin/Stone; B69 (i) T. Eggers/The Stock Market; B70 David Muench Photography, Inc.; B70 (i) Joseph Van Os/The Image Bank; B71 Carr Clifton/Minden Pictures; B71 (i) Kennan Ward Photography; B72 (l) Nicholas DeVore, III/Bruce Coleman, Inc.; B72 (r) Tui De Roy/Minden Pictures; B74 Stan Osolinski/The Stock Market; B80 (t) Jim Brandenburg/Minden Pictures; B80 (b) David Muench Photography, Inc.; B82 (t) © Corel; B82-B83 Manfred Kage/Peter Arnold; B83 (t) NASA GSFC/Science Photo Library/Photo Researchers; B83 (bi) Pete Saloutos/The Stock Market; B84 (t) Romberg Tiburon Center; B84 (b) Emory Kristof/NGS Image Collection; B86 Jim Bradenburg/Minden Pictures; B88-B89 Gary Brettnacher/Stone; B89 (t) Jonathan Wallen; B89 (b) Argus Fotoarchiv/Peter Arnold, Inc.; B90 Frans Lanting/Minden Pictures; B92 Runk/Schoenberger/Grant Heilman Photography; B93 (t) Kennan Ward Photography; B93 (b) Ed Reschke/Peter Arnold, Inc.; B94 (t) Larry Nielsen/Peter Arnold, Inc.; B94 (c) John Marshall/Stone; B94 (b) Jeff & Alexa Henry/Peter Arnold, Inc.; B96 Art Wolfe/Stone; B98 Mark E. Gibson; B98 (i) Dr. E.R. Degginger/Color-Pic; B99 J.H. Robinson/Photo Researchers; B100 Francois Gohier/Photo Researchers; B101 Tony Arruza/Bruce Coleman, Inc.; B104 (c) Jim Corwin/Stone; B106 Tim Davis/Photo Researchers; B110 Mark E. Gibson; B111 (l) Bernard Boutrit/Woodfin Camp & Associates; B111 (r) Bill Tiernan/The Virginian-Pilot; B112 Courtesy of Atlanta Botanical Gardens; B112 (i) Kenneth Murray/Photo Researchers; B114 John Hyde/Bruce Coleman, Inc.; B114 (tli) Superstock; B114 (tri) Tom Bean/The Stock Market; B116 Centre For Ecological Studies; B116 (i) E. Hanumantha/Photo Researchers; B120 (t) Bill M. Campbell, MD; B120 (b) Graeme Teague Photography;

Unit C
Unit C Opener (fg) Eric & David Hosking/Photo Researchers; (bg) Bios (Klein-Hubert)/Peter Arnold, Inc.; C2-C3 Roger Werth/Woodfin Camp & Associates; C3 (t) John Livzey/Stone; C3 (b) Royal Osvervatory, Edinburgh/Science Photo Library/Photo Researchers; C4 Tom Bean/Tom & Susan Bean, Inc.; C6 (tl) Helen Paraskevas; C6 (tr) Tom Bean/Tom & Susan Bean, Inc.; C6 (bi) Mark E. Gibson; C6-C7 Eric Neurath/Stock, Boston; C7 (t) NASA Photo/Grant Heilman Photography; C7 (b) Digital Visual Library/US Army Corps of Engineers; C8 (both) Mark E. Gibson; C9 M.T. O'Keefe/Bruce Coleman, Inc.; C10-C11 Michael Collier/Stock, Boston; C12 Soames Summerhays/Photo Researchers; C16 G. Gualco/Bruce Coleman, Inc.; C17 (t) Gregory G. Dimijian/Photo Researchers; C17 (c) Krafft/Explorer/Science Source/Photo Researchers; C17 (b) Tom & Pat Leeson/Photo Researchers; C18 UPI/Corbis-Bettmann; C20 M.P.L. Fogden/Bruce Coleman, Inc.; C23 Tom Bean/Tom & Susan Bean, Inc.; C24 A. J. Copley/Visuals Unlimited; C25 (t) R.T. Nowitz/Photo Researchers; C26 NASA; C27 (t) Walter H. F. Smith & David T. Sandwell/NOAA National Data Centers; C27 (b) David Young-Wolff/PhotoEdit; C28 (i) Santa Fabio/Black Star/Harcourt; C28 Tom Bean/Tom & Susan Bean, Inc.; C31 (l) Dr. E. R. Degginger/Color-Pic; C31 (r) Joyce Photographics/Photo Researchers; C32-C33 Dan Suzio/Photo Researchers; C33 (tl) Sam Ogden/Science Photo Library/Photo Researchers; C33 (br) Breck P. Kent/Earth Scenes; C34 The Natural History Museum, London; C36 (tl) Dr. E.R. Degginger/Color-Pic; C36 (bl) E.R. Degginger/Bruce Coleman, Inc.; C36 (bc) Joy Spurr/Bruce Coleman, Inc.; C36 (br), (cl) Dr. E.R. Degginger/Color-Pic; C37 (c2), (c3) E.R. Degginger/Bruce Coleman, Inc.; C37 (c5), (c6), (c8) Dr. E.R. Degginger/Color-Pic; C37 (c9) Mark A. Schneider/Dembinsky Photo Associates; C37 (c10) Dr. E.R. Degginger/Bruce Coleman, Inc.; C38 (tl) Dr. E.R. Degginger/Color-Pic; C38 (cl) Biophoto Associates/Photo Researchers; C38 (cr) Andy Sacks/Stone; C38 (bl) Dr. E.R. Degginger/Color-Pic; C38 (br) B. Daemmrich/The Image Works; C40 Joe McDonald/Bruce Coleman, Inc.; C42 (t) Dr. E.R. Degginger/Color-Pic; C42 (b) Phillip Hayson/Photo Researchers; C43 (tl), (tcl) Dr. E.R. Degginger/Color-Pic; C43

(tcr) Breck P. Kent/Earth Scenes; C43 (tr) Robert Pettit/Dembinsky Photo Associates; C43 (b) Martha McBride/Unicorn Stock Photos; C44, C45 (tl), (tcl), (tcr), (tr) Dr. E.R. Degginger/Color-Pic; C45 (b) David Bassett/Stone; C46 (t) G. R. Roberts Photo Library; C46 (b), C46-C47, C47 Dr. E.R. Degginger/Color-Pic; C48 Tom Till/Auscape; C50, C51 (t), (b), C52 (l), (r), C52-C53 Dr. E.R. Degginger/Color-Pic; C54 James P. Blair & Victor Boswell/NGS Image Collection; C55 Mark Richards/Photo Edit; C56 Stuart McCall/Tony Stone Images; C56 (i) Photo Courtesy of Mrs. Alma G. Gipson; C60-C61 Bob Abraham/The Stock Market; C61 (t) NASA/The Stock Market; C61 (b) Stan Osolinski/The Stock Market; C62 (b) Warren Faidley/International Stock Photography; C64 (l) Everett Johnson/Stone; C64 (r) Warren Faidley/International Stock Photography; C65 (bg) Orion/International Stock Photography; C65 (bli) M. Antman/The Image Works; C65 (bri) Dr. E. R. Degginger/Color-Pic; C66 (t) David M. Grossman/Photo Researchers; C66 (b) Mark Stephenson/Westlight; C68 (l) Dan Sudia/Photo Researchers; C68 (tr) Kent Wood/Photo Researchers; C68 (cr) Kevin Schafer/Peter Arnold, Inc.; C68 (br) Gary Meszaros/Dembinsky Photo Associates; C74 Larry Mishkar/Dembinsky Photo Associates; C78 (b) Richard Brown/Stone; C80 (l) Tom Till; C80 (c) Blaine Harrington III/The Stock Market; C80 (r) Coco McCoy/Rainbow; C81 (t) Randy Ury/The Stock Market; C81 (c) Larry Cameron/Photo Researchers; C81 (b) Jeff Greenberg/Photo Researchers; C82 (l) Ron Sefton/Bruce Coleman, Inc.; C82 (cl) Fritz Prenzel/Peter Arnold, Inc.; C82 (c) John Lawrence/Stone; C82 (cr) Marcello Bertinetti/Photo Researchers; C82 (r) Jose Fuste Raga/The Stock Market; C83 Paul Sequeira/Photo Researchers; C83 (i) Joe Sohm/Chromosomm/Photo Researchers; C84-C85 J. Richardson/Westlight; C86 (l) Brad Gaber/The Stock Market; C86-C87 A. Ramey/Woodfin Camp & Associates; C87 (l) NASA; C87 (r) Phil Degginger/Bruce Coleman, Inc.; C88 NASA/Goddard Space Flight Center/Science Photo Library/Photo Researchers; C88 (i) Eli Reichman/HRW; C92-C93 RKO Radio Pictures/Archive Photos; C93 (t) Thomas Abercrombiengs/NGS Image Collection; C93 (b) Clyde H. Smith/Peter Arnold, Inc.; C94 Sylvia Stevens; C97 (t) Timothy O' Keefe/Bruce Coleman, Inc.; C97 (b) Joseph J. Scherschel/NGS Image Collection; C97 (i) Carlos Lacamara/NGS Image Collection; C100 George D. Lepp/Photo Researchers; C102 Vince Cavataio/Pacific Stock; C103 (t) The Stock Market; C103 (ti) Michael P. Gadomski/Bruce Coleman, Inc.; C103 (bl) UPI/Corbis-Bettmann; C103 (br) Chip Porter/AllStock/PNI; C104 (t) Tony Arruza/Bruce Coleman, Inc.; C104 (b) Dr. Richard Legeckis/Science Photo Library/Photo Researchers; C105 George Marler/Bruce Coleman, Inc.; C106 (l), (r) John Elk/Stock, Boston; C110 (l) Brian Parker/Tom Stack & Associates; C110 (r) S.L. Craig, Jr./Bruce Coleman, Inc.; C111 (l) William E. Ferguson; C111 (r) Toms & Susan Bean, Inc.; C112 (t) Bruce Roberts/Photo Researchers; C112 (c) William Johnson/Stock, Boston; C112 (b) Wendel Metzer/Bruce Coleman, Inc.; C112-C113 Bob Daemmrich/Stock, Boston; C114 Michael Paris Photography; C116 (t) The Granger Collection, New York; C116 (bl) Corbis-Bettmann; C116 (br) The Granger Collection, New York; C116-C117 Eric Le Norcy-Bios/Peter Arnold, Inc.; C117 (tl) Naval Undersea Museum; C117 (tr) NASA/Science Photo Library/Photo Researchers; C117 (bl) Courtesy Smithsonian Diving Office/Photograph by Diane L. Nordeck; C117 (br) James P. Blair/NGS Image Collection; C118 (t) Woods Hole Oceanographic Institution; C118 (bl), (br), C119 (t) Emory Kristof/NGS Image Collection; C119 (c) Culver Pictures; C119 (b) Woods Hole Oceanographic Institution; C120 Allen Green/Photo Researchers; C121 Christian Vioujard/Gamma Liaison; C122 (t) © PhotoDisc; C122 (b) Mike Price/Bruce Coleman, Inc.; C123 (t) Eric Freedman/Bruce Coleman, Inc.; C123 (b) David Woodfall/Stone; C124 Susan Lapides/Woodfin Camp & Associates; C124 (i) Emory Kristof/NGS Image Collection; C128 (t) Brownie Harris/The Stock Market; C128 (b) Adam Jones/Dembinsky Photo Associates;

Unit D
Unit D Opener (fg), (bg) Telegraph Colour Library/FPG International; D2-D3 Guodo Cozzi/Bruce Coleman, Inc.; D3 (l) Ray Pfortner/Peter Arnold, Inc.; D3 (r) Painting by Helmut Wimmer; D4 NASA; D6, D7 Frank Rossotto/StockTrek; D8 (t) Dennis Di Cico/Peter Arnold, Inc.; D8 (b) Frank Rossotto/StockTrek; D9, D10 (tr), (tl) NASA; D10 (ctl) Francois Gohier/Photo Researchers; D10 (ctr), (cbr) NASA; D10 (bl) Paul Stepan/Photo Researchers; D10 (br), D12 NASA; D14 (l) Peter Marbach; D14 (r) Jeff Greenberg/Unicorn Stock Photos; D15 Fred Habegger/Grant Heilman Photography; D18 NASA; D18 (Pluto) Dr. R. Albrecht, ESA/ESO Space Telescope European Coordinating Facility/NASA; D19 E. R. Degginger/Color-Pic; D20 NASA; D22 (t) The Granger Collection, New York; D22 (bl) Martha Cooper/Peter Arnold, Inc.; D22 (br) The Granger Collection, New York; D22, D23, D24 (bg) Science Photo Library/Photo Researchers; D23 (tl) Sovfoto/Eastfoto; D23 (tr) NASA; D23 (bl) Courtesy of AT&T Archives; D23 (br),D24, D25, D26-D27, D28, D29, D30, D33 NASA; D34-D35 Jerry Schad/Photo Researchers; D35 NASA; D36 StockTrek; D39 (t) Warren Faidley/International Stock Photography; D39 (c) Pekka Parviainen/Science Photo Library/Photo Researchers; D39 (b) Brian Atkinson/Stone; D40 (t) Rev. Ronald Royer/Science Photo Library/Photo Researchers; D40 (c) NASA; D40 (b) Hale Observatory/SS/Photo Researchers; D42 Wards Sci/Science Source/Photo Researchers; D46 John Chumack/Photo Researchers; D48 Andrea Dupree (Harvard-Smithsonian CfA), Ronald Gilliland (STScI), NASA and ESA; D49 Jeff Hester and Paul Scowen (Arizona State University), and NASA; D50 (l) NASA; D50 (r) Roger Ressmeyer/Corbis; D50 (b) Francois Gohier/Photo Researchers; D52 Bill Ross/Stone; D54 Fred Espenak/Science Photo Library/Photo Researchers; D55 Lynette Cook/Science Photo Library/Photo Researchers; D56 (t) Royal Observatory, Edinburgh/AATB/Science Photo Library/Photo Researchers; D56 (b) Robert Williams and the Hubble Deep Field Team (STScI) and NASA; D58 Dr. Robert Mallozzi of Universtiy of Alabama/Huntsville & NASA; D58-D59 Chris Cheadle/Stone; D59 Tony Freeman/PhotoEdit; D60 Courtesy of Julio Navarro/University of Victoria; D63 (l) Lynette Cook/Science Photo Library/Photo Researchers; D64 (t) Andre Jenny/Unicorn Stock Photos; D64 (b) Dennis Johnson/Folio;

Unit E
Unit E Opener (fg) Alvis Upitis/The Image Bank; (bg) Tim Crosby/Liaison International; E2 Jim Steinberg/Photo Researchers; E2-E3 Charles Krebs/The Stock Market; E3 H. Armstrong Roberts; E7 (b) NASA; E10 (bg) Ron Chapple/FPG International; E10 (bi) Dr. E. R. Degginger/Color-Pic; E12 Charles D. Winters/Photo Researchers; E14 Spencer Swanger/Tom Stack & Associates; E15 (c) Phil Degginger/Color-Pic; E16 Dr. E. R. Degginger/Color-Pic; E16-E17 Tom Pantages; E17 Tom Pantages; E18 Yoav Levy/Phototake; E19 Yoav Levy/Phototake; E20 NASA; E23 (t) Tom Pantages; E23 (c) Yoav Levy/Phototake; E23 (b) Tom Pantages; E24 Horst Desterwinter/International Stock Photography; E25 (t) Dr. E.R. Degginger/Color-Pic; E25 (i) Norman O. Tomalin/Bruce Coleman, Inc.; E28 Joe Sohm/Photo Researchers; E29 (t) Doug Martin/Photo Researchers; E29 (b) Gary A. Conner/PhotoEdit; E30 (i) Glenn Photography; E30 Geoff Tompkinson/Science Photo Library/Photo Researchers; E34 Jan Taylor/Bruce Coleman, Inc.; E34-E35 Pete Saloutos/Stone; E35 Dr. E.R. Degginger/Color-Pic; E36 Superstock; E38 Lee Snider; E40 (t) J & L Weber/Peter Arnold, Inc.; E40 (b) Dr. E.R. Degginger/Color-Pic; E41 (Top to bottom) Joe Towers/The Stock Market; E41 (photo 2) Christopher S. Johnson/Stock, Boston; E41 (photo 4) Telegraph Colour Library/FPG International; E41 (photo 6) George Haling/Photo Researchers; E42 (li) Richard Laird/FPG International; E42 (b) Wesley Hitt/Stone; E44 Yoav Levy/Phototake; E48 Michael Monello/Julie A. Smith Photography; E50 (t) © PhotoDisc; E50 (bl) Mel Fisher Maritime Heritage Society, Inc.; E50-E51 (bg) Michigan Molecular Institute; E51 Michigan Molecular Institute; E52 (i) UPI/Corbis; E52 Mitch Kezar/Phototake; E56 (t) Sal Dimarco/Black Star/Harcourt; E56 (b) Courtesy of Jefferson Lab;

Unit F
Unit F Opener (fg) David Zaitz/Photonica; (bg) Stone; F2-F3 NASA/Photo Researchers; F3 (t) NASA/ Science Photo Library/Photo Researchers; F3 (b) Jean-Loup Charmet/Science Photo Library/Photo Researchers; F6 (l) Tony Duffy/Allsport Photography; F6 (c) Pascal Rondeau/Allsport Photography; F6 (r) Allsport Photography; F7 (tr) Spencer Grant/PhotoEdit; F7 (cr) Spencer Grant/PhotoEdit; F10 (b) David Young-Wolff/PhotoEdit; F12 (l) Myrleen Ferguson/PhotoEdit; F12 (r) Mark E. Gibson; F18 Superstock; F19 David Young-Wolff/PhotoEdit; F20 (b) Stephen Frisch/Stock, Boston; F21 (t) Novastock/PhotoEdit; F21 (ct) Tony Freeman/PhotoEdit; F21 (b) Tony Freeman/PhotoEdit; F24 NASA; F25 NASA; F26 (i) Dr. Ephraim Fischbach; F26 NASA; F30-F31 Chris Butler/Science Photo Library/Photo Researchers; F31 (t) Stephen Dalton/Photo Researchers; F31 (b) Mike Cooper/Allsport Photography; F32 (bl) E & P Bauer/Bruce Coleman, Inc.; F35 David Madison/Bruce Coleman, Inc.; F36 (i) Mark E. Gibson; F36-F37 Lee Foster/Bruce Coleman, Inc.; F38 (bl) Michael Newman/PhotoEdit; F40 A. C. Cooper LTD/Harcourt; F40-F41 Ed Degginger/Bruce Coleman, Inc.; F41 NASA; F42 Tony Freeman/PhotoEdit; F43 Mike Yamashita/The Stock Shop; F44 CP Picture Archive (Chuck Stoody); F46 Tom McHugh/Photo Researchers; F49 Erich Lessing/Art Resource, NY; F51 Scala/Art Resource, NY; F52-F53 Lawrence Livermore National Laboratory; F53 (t) Lawrence Livermore National Laboratory; F53 (b) Michael Rosenfeld/Stone; F54 NASA; F58-F59 Scott Warren; F59 (b) M. Zhilin/M. Newman/Photo Researchers; F60 Jan Butchofsky-Houser/Dave G. Houser; F62 (l) Duomo Photography; F62 (c) William R. Sallaz/Duomo Photography; F62 (r) Steven E. Sutton/Duomo Photography; F63 (t) Gary Bigham/International Stock Photography; F63 (ti) Ken Gallard Photographics; F63 (bl) Stevn E. Sutton/Duomo Photography; F63 (br) Duomo Photography; F64 Greg L. Ryan & Sally Beyer/All-Stock/PNI; F66 Peter Menzel; F68-F69 Phil Degginger/Bruce Coleman, Inc.; F69 (tr) Ontario Science Centre; F69 (br) Phil Degginger/Bruce Coleman, Inc.; F72 (tr) Michael J. Schimpf; F74 (bl) Stone; F76 (l) E.R. Degginger/Bruce Coleman, Inc.; F76 (c) E.R. Degginger/Bruce Coleman, Inc.; F76 (r) Tony Freeman/PhotoEdit; F78 (tl) Tim Beddow/Stone; F79 (tr) Danila G. Donadoni/Bruce Coleman, Inc.; F80 Norbert Wu/Stone; F82 Chuck O'Rear/H. Armstrong Roberts, Inc.; F86-F87 Peter Cade/Stone; F88 (t) Michael Keller/The Stock Market; F88 (bl) The Granger Collection; F88 (bc) Kevin Collins/Visuals Unlimited; F88 (br) American Stock Photography; F89 (tl) Corbis-Bettmann; F89 (bl) Michael Nelson/FPG International; F89 (r) Lester Lefkowitz/The Stock Market; F90 (i) Dr. Jean M. Bennett; F90 Diane Schiumo/Fundamental Photographs; F94-F95 Jeff Hunter/ The Image Bank; F95 (t) Science Photo Library/Photo Researchers; F95 (bl) James King-Holmes/ Science Photo Library/Photo Researchers; F95 (br) Ron Chapple/FPG International; F96 Cary Wolinsky/Stock, Boston/PNI; F98 Billy E. Barnes/PhotoEdit; F99 (tl) Gary Conner/PhotoEdit; F99 (tr) Bill Aron/PhotoEdit; F99 (br) Myrleen Ferguson/PhotoEdit; F100 (tl) Tony Freeman/PhotoEdit/PNI; F100 (br) David Young-Wolfe/PhotoEdit; F102 Jim McCrary/Stone; F104 (b) Wendell Metzen/Bruce Coleman, Inc.; F104 (i) Mark E. Gibson; F105 (tl) Keith Gunnar/Bruce Coleman, Inc.; F105 (tr) Mark Newman/Bruce Coleman, Inc.; F105 (c) Tony Freeman/PhotoEdit; F106 (c) Claus Militz/Okapia/Photo Researchers; F108 (bl) Myrleen Ferguson/PhotoEdit; F110 (t) Alan L. Detrick/Photo Researchers; F110 (b) Cameramann International; F111 (t) Nicholas de Vore III/Bruce Coleman, Inc.; F111 (c) Andrew Rakoczy/Bruce Coleman, Inc.; F111 (b) Glen Allison/Stone; F112 (tr) Lockheed Space and Missile Co., Inc.; F114 (t) © Corel; F114-F115 Mark E. Gibson; F115 (t) M.E. Rzucidlo/H. Armstrong Roberts; F115 (b) Andy Sacks/Stone; F116 (l) Drew Donovan Photography; F116 (r) Stan Ries/International Stock Photography; F120 (t) Courtesy of the National Science Center's Fort Discovery; F120 (b) Courtesy Associated Electric Cooperative, Inc.;

Health Handbook
R16 (c) Tony Freeman/PhotoEdit; R16 (br) David Young-Wolff/PhotoEdit; R17 (t) Myrleen Ferguson Cate/PhotoEdit; R17 (b) David Young-Wolff/PhotoEdit; R19 Tony Freeman/PhotoEdit; R45 (br) © PhotoDisc; R45 (bl) PhotoDisc;

All other photos by Harcourt photographer listed below, © Harcourt: Weronica Ankarorn, Victoria Bowen, Eric Camden, Digital Imaging Gorup, Charles Hodges, Ken Kinzie, Ed McDonald, Sheri O'Neal, Terry Sinclair & Quebecor Digital Imaging.

Illustration credits - Tim Alt B14-15, E15, E16-17, E39, E40, E41, F84; Scott Angle R44, R45; Art Staff C66; Paul Breeden A94; Lewis Calver A8, A9, A10, A11, A12, F78-79; Mike Dammer A57, A117, B85, C29, C57, C89, C125, D31, E53, F55, F117; John Dawson B38, B78-79; Eldon Doty A31, A57, A117, B57, C29, C125, D31, E53, F55, F117; Pat Foss A85, A117, B21, B57, B85, B117, E31, F27, F91; George Fryer C44, C46; Patrick Gnan F22, F65, F77, F78; Dale Gustafson B7, F64, F70-71, F72, F106; Nick Hall C102, C106; Tim Hayward A40; Jackie Heda A16, A17, A18, A19, A20, A25, A26; Inklink F20-21; Roger Kent A95, A100, A101, B44; Mike Lamble C16, C17, D6-7, D8, D9, D33, D55, F34-35, F48-49, F50; Ruth Lindsay B8-9, B92-93, B106-107; Lee MacLeod A27; Alan Male A79, A80, B43; MapQuest C75, C96; Janos Marffy C22, D15; Colin Newman B76-77, C110-111; Sebastian Quigley C64-65, C67, C72, C73, C75, C76, D14, D16-17, D38-39, D40-41, D46, D47, D48-49, F7; Rosie Sanders A102, A103; Mike Saunders A50, A51, C9, C10, C14-15, C18-19, C24 ,C31, C50-51, C98-99, C105, C118-119, F8; Andrew Shiff A31, B21, B117, C87, D61, E31, F27, F91; Steve Westin C42-43, C45; Beth Willert A24.